Methods of
Digital Holography

Methods of
Digital Holography

L. P. Yaroslavskii
and
N. S. Merzlyakov

Academy of Sciences of the USSR
Moscow, USSR

Translated from Russian by
Dave Parsons

CONSULTANTS BUREAU · NEW YORK AND LONDON

Library of Congress Cataloging in Publication Data

IAroslavskii, Leonid Pinkhusovich.
  Methods of digital holography.

  Translation of Metody tsifrovoi golografii.
  Includes index.
  1. Holography—Data processing. I. Merzliakov, Nikolai Stepanovich, joint author.
II. Title.
TA1542.12713   621.36'75'02854   80-16286
ISBN 0-306-10963-8

© 1980 Consultants Bureau, New York
A Division of Plenum Publishing Corporation
227 West 17th Street, New York, N.Y. 10011

Printed in the United States of America

Preface to the English Edition

We are very pleased that *Methods of Digital Holography* is being published in English translation in the USA, where the earliest work on digital holography was carried out and where many outstanding results have been obtained. To the best of our knowledge, this translation will become the first book in English devoted solely to digital holography, so we feel a special responsibility. We have taken the opportunity presented by the publication of this translation to correct some minor errors in the Russian edition, to add some new and better illustrations, and to extend the list of references slightly. Unfortunately, it was not possible to make more substantial additions.

Digital holography belongs both to applied physics and computer science, and we hope that this book will contribute to both of these broad areas of research, which stand at the frontier of technological progress. We also hope that the book will help demonstrate that digital holography constitutes a research field in its own right, with its own methods and goals.

We would like to thank A. Lohmann of the Physikalisches Institut der Universität, Erlangen, FRG, and G. W. Stroke, of the State University of New York at Stony Brook, for some useful and stimulating discussions of this book.

L. P. Yaroslavskii
N. S. Merzlyakov

Preface

Digital holography is the analysis and synthesis of wavefronts by means of digital computers. Here "analysis" means the construction of images of objects and the measurement of their physical characteristics by recording and measuring the wavefronts scattered by these objects. The "synthesis" is the reconstruction of a wavefront of an object which is specified numerically.

Problems involving the analysis and synthesis of wavefronts arise frequently in modern science and engineering. These are the problems encountered in the use of electromagnetic and sound waves to peer into the interior of various objects, the problems of the visual display of information in general, and the problems of measuring the characteristics of radiating systems, devising optical apparatus for signal processing, and developing hybrid electrooptic computer systems. Solutions are being sought for these problems through research in holography.

The use of digital computers to analyze and synthesize wavefronts is an alternative to the analog methods, which include the methods of physical holography. The digital approach potentially has the advantages which are inherent in the digital technique for signal processing: the processing is highly accurate and absolutely reproducible; the characteristics of the processing or the processing algorithm itself can be changed in a simple way; complicated nonlinear and logic conversions can be carried out; the results are accessible; and it is easy to modify the process at any stage. Digital methods are particularly suitable where quantitative results are required.

In order actually to realize these potential capabilities, we need to develop the appropriate methods and apparatus for digital holography.

The purpose of the present book, which is based primarily on the experience acquired by the authors in their work on digital holography, is to examine the basic problems of digital holography from a common standpoint. These problems are those of synthesizing the holograms, analyzing them (carrying out the reconstruction), and developing the appropriate hardware and software.

These three problems of digital holography are taken up in the three chapters of this book. The first deals primarily with the use of synthesized holograms to visualize objects which are specified numerically. This chapter covers the mathematical models which are used to synthesize the holograms, the theory of fast Fourier transform algorithms, the characteristics of analog devices for recording holograms, the effect of these characteristics on the resulting images, the results of experiments with synthesized holograms, the results of experiments with stereoholograms, and the computer synthesis of three-dimensional holographic film and holograms with a programmable diffuser.

The second chapter covers the analysis of holograms for forming images and the measurement of patterns of radiating systems. In this chapter we also describe a digital model for studying the "speckle noise" which arises in reconstruction from the holograms.

The third chapter describes two devices for entering holograms in computers and for recording the holograms generated by computers; discusses apparatus for photoenlargement, photoreduction, and reconstruction of holograms; and describes program libraries for synthesizing and analyzing holograms.

Reference material on the properties of one- and two-dimensional discrete Fourier transformations is included in Appendixes I and II.

Sections 1–3, 7, and 8 and Appendixes I and II were written by L. P. Yaroslavskii. Sections 4–6 and Chapter 3 were written by the authors jointly. The experimental results discussed in Section 8 were furnished by N. R. Popova. The authors thank V. N. Karnaukhov for assistance in the experiments on the reconstruction from synthesized holograms and the development of a holographic film.

Contents

1

Synthesis of Holograms

The invention of optical holography [1-14] furnished a radically new solution to the problem of preserving video information on three-dimensional objects and of visualizing this information. In principle, a hologram records the distribution of the complex amplitude of the wavefront scattered by an object so that reconstruction from the hologram produces a complete visual illusion of the object. To observe the image of an object produced from a hologram is nearly as convenient and natural as to observe the object itself.

There are many problems, however, in which a three-dimensional visualization is required of objects which are very difficult or impossible to reproduce, for one reason or another, but for which an accurate mathematical description can be found. Examples are the visualization of information from a computer and the development of computer-controlled or programmed transparencies.

A question which arises here is whether these problems can be solved by using holographic methods to record video information—by synthesizing holograms in accordance with a numerical description of the object.

An analogous problem arises in the development of three-dimensional television. The current research in this field is based on a holographic approach. It may turn out that the synthesis of holograms at the receiver will open the door to holographic television.

In this chapter we will report a study of the computer synthesis of holograms for visualizing information.

The hologram-synthesis problem is formulated as follows: There exists a mathematical description of an object such that the amplitude and phase of the light in it can be found for an arbitrary or specified position of the light source. The position of the observer with respect to the object is also specified. We are required to calculate the amplitude and phase of the light scattered by the object at each observation point and to record the calculated results in the form of a physical hologram from which it would be possible to observe the original object from the desired perspective and with suitable illumination. The solution of this problem requires the following steps:

1. The development of a discrete representation of those transformations which relate the amplitude and phase distributions of the light at the object to the corresponding distributions at the observation points.
2. The development of algorithms and programs to implement step 1.
3. The development of the technical means for recording synthesized holograms which are capable of producing a physical hologram from information supplied by a computer.
4. The development of reconstruction and observation methods suitable for use with synthesized holograms.

Section 1. Mathematical Model

Mathematical Hologram. In order to choose a formal apparatus to solve the problem of synthesizing holograms for visualizing objects which are specified by some mathematical description, and to understand what is required of this description, we will consider the scheme arrangement in Fig. 1.1 for the visual observation of objects. The position of the observer with respect to the object is described by an observation surface on which the observer's eye lies; the set of perspectives is described by the "coverage" angle. The observer will be able to see the object in the given coverage angle if the intensity and phase distribution of the light wave scattered by the object is reproduced at the observation surface by means of a hologram.

Those characteristics of the object which determine its ability to reflect and scatter incident light are described by the intensity reflectance $B(x, y, z)$ or by the amplitude reflection coefficient $b(x, y, z)$, both of which are functions of

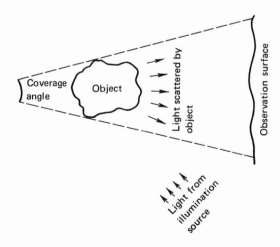

Fig. 1.1

the coordinates on the surface of the object.† The amplitude reflection coefficient is a complex function which can be written

$$b\,(x,\,y,\,z) = |\,b\,(x,\,y,\,z)\,|\,\exp\,[i\beta\,(x,\,y,\,z)].\tag{1.1}$$

The modulus $|b|$ and phase (β) of this coefficient show the factor by which the amplitude is changed and, correspondingly, the extent to which the phase of the light is changed at the point (x, y, z) on the surface of the object after reflection. The functions B and b are related by

$$B = |b|^2 = bb^*,\tag{1.2}$$

where the asterisk denotes the complex conjugate. Knowing the function $b(x, y, z)$, the equation of the surface of the object $[F(x, y, z) = 0]$, and the amplitude and phase distribution of the light incident on the sample, we can in principle calculate the amplitude and phase distribution of the light at the observation surface.

Denoting by $A\,(x, y, z)\,\exp\,i\alpha(x, y, z)$ the amplitude and phase distribution of the light at the surface of the object, we can describe the field at the observation surface by the integral relation

$$\Gamma\,(\xi,\,\eta,\,\zeta) = \int_{F(x,\,y,\,z)} A\,(x,\,y,\,z)\,|\,b\,(x,\,y,\,z)\,| \times$$
$$\times \exp\,i\,[\alpha\,(x,\,y,\,z) + \beta\,(x,\,y,\,z)] \times$$
$$\times\,T\,(x,\,y,\,z,\,\xi,\,\eta,\,\zeta)\,dx\,dy\,dz,\tag{1.3}$$

where the integration is carried out over the surface $F(x, y, z)$. The kernel of this transformation, $T(x, y, z, \xi, \eta, \zeta)$, depends on the spatial positions of the object and the observation surface. In principle, this transformation is invertible:

$$b\,(x,\,y,\,z) = \int_{S} \Gamma\,(\xi,\,\eta,\,\zeta)\,\overline{T}\,(x,\,y,\,z,\,\xi,\,\eta,\,\zeta)\,d\xi\,d\eta\,d\zeta,\tag{1.4}$$

where $\overline{T}$ is the operator which is the inverse of T, and the integration is carried out over the observation surface, S.

The function $\Gamma(\xi, \eta, \zeta)$ may be called the "mathematical hologram." The hologram-synthesis problem is one of calculating the function $\Gamma(\xi, \eta, \zeta)$ from a given function $b(x, y, z)$ and of recording the result in a form which would allow an interaction with light for visualization or reconstruction of $b(x, y, z)$ in accordance with (1.4).

---

†For simplicity, we are restricting the discussion to monochromatic illumination and to opaque objects which emit no light of their own.

The evaluation of the integral in Eq. (1.3) in the general case, however, is a complicated problem.

The first simplification to which we can resort without essentially changing the problem is to reduce the three-dimensional problem to a two-dimensional problem. For this purpose, we assume that the observation surface is a plane, and we use the laws of geometric optics to replace the amplitude and phase distribution of the wave on the object surface by the amplitude and phase distribution on a plane tangent to the object (or on a plane which is close enough to the object so that we can ignore diffraction in the conversion of the wave amplitude and phase and so that we can use geometric optics) and parallel to the observation plane.

Then we can replace Eq. (1.3) by

$$\Gamma(\xi,\eta) \sim \int\limits_{(x,\,y)} b_1(x,\,y)\,T_d(x,\,y,\,\xi,\,\eta)\,dx\,dy, \qquad (1.5)$$

where $b_1(x,y)$ is the complex function which is found by referring the amplitude and phase of the field reflected by the object to the (x,y) plane, which is tangent to the object and parallel to the observation plane, (ξ,η), and d is the distance between these two planes (Fig. 1.2).

This approximation is obviously a natural one if the coverage angle and the observation area are both small. For problems in which the coverage angle must be large, this approach means that the problem must be reduced to a visualization problem with a small coverage angle. Large coverage angles could also be arranged by synthesizing composite holograms, each part of which represents a fraction of the total angle.

The kernel of the transformation in (1.5), which relates the light distributions on the two parallel planes, is

$$T_d(x,\,y,\,\xi,\,\eta) = \frac{\exp\left(i2\pi\lambda^{-1}\sqrt{(x-\xi)^2 + (y-\eta)^2 + d^2}\right)}{\sqrt{(x-\xi)^2 + (y-\eta)^2 + d^2}}, \qquad (1.6)$$

where λ is the wavelength of the light.

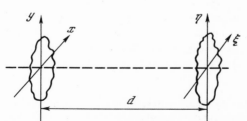

Fig. 1.2

If the geometric dimensions of the object are small in comparison with d, which is the distance to the observation plane (and if the observation area is small), we find a further simplification:

$$T_d(x, y, \xi, \eta) \simeq \{\exp(i2\pi\lambda^{-1} d) \exp\{i\pi(\lambda d)^{-1} \times \\ \times [(x-\xi)^2 + (y-\eta)^2]\}\}/d \tag{1.7}$$

for

$$[(x-\xi)^2 + (y-\eta)^2]_{\max}/d^2 = \theta^2_{\max} < \sqrt{\frac{4\lambda}{kd}}, \tag{1.8}$$

where $\theta_{\max}$ is the maximum angle (in radians) at which the object is observed from the distance d, and k is the coefficient of the permissible phase error (the error is π/k) in the propagation of the argument of the exponential function in (1.6).

In this case the integral in (1.5) converts into a Fresnel integral:

$$\Gamma(\xi, \eta) \sim \int_{(x, y)} b_1(x, y) \exp\{i\pi\lambda^{-1} d^{-1} \times \\ \times [(x-\xi)^2 + (y-\eta)^2]\} \, dx \, dy. \tag{1.9}$$

Those holograms which are synthesized in accordance with this relation will be called "synthesized Fresnel holograms."

A further simplification is possible if

$$\pi(x^2 + y^2)/\lambda d \ll 1, \quad \pi(\xi^2 + \eta^2)/\lambda d \ll 1, \tag{1.10}$$

so that these phase components can be ignored in the integral in (1.9). Then (1.9) converts into a Fourier integral,†

$$\Gamma(\xi, \eta) \sim \int_{(x, y)} b_1(x, y) \exp[i2\pi(\lambda d)^{-1}(x\xi + y\eta)] \, dx \, dy. \tag{1.11}$$

This case is of particular importance for optical information-processing systems because the amplitude of the Fourier transform is invariant with respect to displacements along the coordinates; thus convolution operations can be carried out in optical systems.

---

†This case corresponds to Fraunhofer diffraction. Equation (1.11) also describes the properties of the lens as an optical element which performs a Fourier transformation.

We will label those holograms which are synthesized in accordance with (1.11) "synthesized Fourier holograms" or simply "Fourier holograms." It follows from (1.11) that a Fourier hologram is, within unimportant amplitude factors, the spatial Fourier spectrum of the function $b_1(x, y)$

$$F_b(v_x, v_y) = \iint\limits_{(x, y)} b_1(x, y) \exp[i2\pi(xv_x + yv_y)] \, dx \, dy, \qquad (1.12)$$

taken along the coordinates v_x, v_y in the scale $v_x = \xi/\lambda d, v_y = \eta/\lambda d$,

$$= \eta/\lambda d,$$
$$\Gamma(\xi, \eta) = F_b(v_x = \xi/\lambda d, \ v_y = \eta/\lambda d). \qquad (1.13)$$

The Fresnel transformation can be thought of as the convolution of the object with a pulsed response as in (1.7). From the computational standpoint, however, it is more convenient to express (1.11) in terms of a Fourier transformation:

$$\Gamma_0(\xi, \eta) = \exp[i\pi(\lambda d)^{-1}(\xi^2 + \eta^2)] \iint\limits_{(x, y)} b_1(x, y) \times$$
$$\times \exp[i\pi(\lambda d)^{-1}(x^2 + y^2)] \times$$
$$\times \exp[-i2\pi(\lambda d)^{-1}(x\xi + y\eta)] \, dx \, dy. \qquad (1.14)$$

According to (1.14), a Fresnel hologram is a Fourier hologram of the same object but observed through a lens; the Fourier hologram itself is also recorded at the exit from the lens. This second lens, which is described by the phase factor $\exp\{i(2\pi/\lambda d)(\xi^2 + \eta^2)\}$ in (1.14), is by no means a mandatory part of the synthesis process since it contains the parameter d, which is the same for the entire hologram and which is furthermore extremely simple. This parameter can be reconstructed separately.†

Finally, we note that by transforming from a three-dimensional problem to a two-dimensional problem we are, strictly speaking, losing the capability of accurately determining the depth of the surface relief of the object. Even the Fresnel hologram contains only the distance from the object to the observation plane—not the depth of the relief. Nevertheless, it is possible to synthesize a field which can reconstruct the object under certain conditions. Then the most important property of holographic visualization—the natural way in which the object is observed—is retained. Certain artificial methods can be suggested for conveying the relief (Section 5).

Representation of a Fourier Hologram in Terms of a Discrete Fourier Transformation. Equation (1.12), with (1.13), is the starting point for the computer synthesis of Fourier holograms. The computer method used for this discrete

---

†We will see in Section 5 that in several cases we can also ignore the first lens.

synthesis depends on the form of the discrete description (digitization) of the amplitude transmission coefficient of the transparency, $b_1(x, y)$. The simplest way to digitize $b_1(x, y)$ is to specify it as a matrix of numbers ("readings") $b_1(k, l)$ taken on a rectangular raster with certain steps Δx and Δy along the coordinates. This description is based most naturally on the sampling theorem. To transform from this discrete description to a continuous description of $b_1(x, y)$, we use a linear interpolation of the readings. Mathematically, the interpolation can be described as the transformation of the series

$$b_1(x, y) = \sum_{k=0}^{N_x-1} \sum_{l=0}^{N_y-1} b_1(k, l)\, \delta(x - k\Delta x, y - l\Delta y) \qquad (1.15)$$

by a continuous linear operator. The ranges of k, l, N_x, and N_y are governed by the dimensions of the transparency and by the digitization step.

If $b_1(x, y)$ is nonvanishing in the rectangle $[-X_{max}, X_{max}; -Y_{max}, Y_{max}]$, then

$$N_x = 2X_{max}/\Delta x, \quad N_y = 2Y_{max}/\Delta y. \qquad (1.16)$$

An accurate interpolation of $b_1(x, y)$ from $b_1(k, l)$ is possible if

$$\Delta x = 1/2 \; v_{xmax}; \quad \Delta y = 1/2 \; v_{ymax}, \qquad (1.17)$$

where $[-v_{xmax}, v_{xmax}; -v_{ymax}, v_{ymax}]$ is the rectangular region within which the spatial spectrum of $b_1(x, y)$ is nonvanishing.

Since calculations can be carried out on computers only for discrete objects, the synthesis of the hologram of the original object $b_1(x, y)$ must be broken up into two steps: a digital synthesis of the hologram of the discrete object, $\hat{b}_1(x, y)$, from the matrix of readings $b_1(k, l)$ and an analog reconstruction of a continuous object in the visualization step.

Substitution of (1.15) into (1.12) shows that the Fourier hologram of the discrete object $b_1(x, y)$ can be calculated as the finite sum

$$F_{\hat{b}}(v_x, v_y) = \sum_{k=0}^{N_x-1} \sum_{l=0}^{N_y-1} b_1(k, l) \exp\left[i2\pi\left(\frac{kv_x}{2v_{x\,max}} + \frac{lv_y}{2v_{y\,max}}\right)\right]. \qquad (1.18)$$

Now we must extend the digitization in (1.18) to v_x and v_y. This procedure can be justified on the basis that the dimensions of the object $b_1(x, y)$ are bounded, or at the very least these dimensions can be assumed bounded within a known error.† It follows from the boundedness of the region in which $b(x, y)$ is specified that $F_b(v_x, v_y)$ and thus, within a known error, $F_{\hat{b}}(v_x, v_y)$ can be reconstructed by a linear interpolation of the discrete function

$$F_{\hat{b}}(v_x, v_y) = \sum_{r=0}^{N_x-1} \sum_{s=0}^{N_y-1} F_{\hat{b}}(r\Delta v_x, s\Delta v_y)\, \delta\,(v_x - r\Delta v_x, v_y - s\Delta v_y), \quad (1.19)$$

where $F_{\hat{b}}(r\Delta_x, s\Delta v_y)$ are the readings of $F_{\hat{b}}(v_x, v_y)$ taken on a rectangular raster with steps Δv_x, Δv_y along the coordinates v_x, v_y, and

$$F_{\hat{b}}(r, s) = \sum_{k=0}^{N_x-1} \sum_{l=0}^{N_y-1} b_1(k,l) \exp\left[i2\pi\left(kr\,\frac{\Delta v_x}{2v_{x\,max}} + ls\,\frac{\Delta v_y}{2v_{y\,max}}\right)\right]. \quad (1.20)$$

The problem of synthesizing a Fourier hologram of the object $b_1(x, y)$ is thus reduced to a numerical calculation of the matrix $F_{\hat{b}}(r, s)$ from the matrix of readings $b_1(k, l)$ and two analog procedures: an interpolation to find $F_{\hat{b}}(v_x, v_y)$ and an interpolation on the basis of the reconstructed readings $b_1(k, l)$. These analog procedures will be discussed below.

For $v_x, v_y, v_{xmax}, v_{ymax}, N_x$, and N_y, the following relations hold:

$$\begin{aligned} M_x &= 2v_{x\,max}/\Delta v_x = N_x = 2X_{max}/\Delta x, \\ M_y &= 2v_{y\,max}/\Delta v_y = N_y = 2Y_{max}/\Delta y. \end{aligned} \quad (1.21)$$

Substitution of (1.21) into (1.20) finally yields an equation for calculating the elements of the matrix $\{F_{\hat{b}}(r, s)\}$ from the matrix of numbers $\{b_1(k, l)\}$

$$F_{\hat{b}}(r, s) = \sum_{k=0}^{N_x-1} \sum_{l=0}^{N_y-1} b_1(k, l) \exp\left[i2\pi\left(\frac{kr}{N_x} + \frac{ls}{N_y}\right)\right]. \quad (1.22)$$

This equation is a two-dimensional discrete Fourier transformation. A discrete Fourier transformation (DFT) is related to an integral Fourier transformation, but it has certain distinct features which stem from the fact that the metric of the argument space is an integer. The properties of the two- and one-dimensional DFTs are examined in Appendixes I and II.

---

†The same stipulation should be made regarding the boundedness of the spatial spectrum of $F_{\hat{b}}(v_x, v_y)$. An analysis of the distortions which arise if this condition is not met goes beyond the scope of the present work, but this subject is dealt with in the theory of the digitization of signals.

A two-dimensional DFT can be calculated as two successive one-dimensional Fourier transformations†:

$$F_{\hat{b}}^0(r, l) = \sum_{k=0}^{N_x-1} b_1(k, l) \exp\left(i2\pi \frac{kr}{N_x}\right),$$ (1.23)

$$F_{\hat{b}}(r, s) = \sum_{l=0}^{N_y-1} F_{\hat{b}}^0(r, l) \exp\left(i2\pi \frac{ls}{N_y}\right).$$ (1.24)

Representation of a Fresnel Hologram in Terms of Discrete Fourier Transformation. We now consider a discrete representation of a Fresnel hologram:

$$\Gamma(\xi, \eta) = \iint_{(x, y)} b_1(x, y) \exp\left\{i \frac{\pi}{\lambda d} [(x - \xi)^2 + (y - \eta)^2]\right\} \times$$
$$\times\, dx\, dy = \exp\left[i \frac{\pi}{\lambda d} (\xi^2 + \eta^2)\right] \iint_{(x, y)} b_1(x, y) \times$$
$$\times \exp\left[i \frac{\pi}{\lambda d} (x^2 + y^2)\right] \exp\left[-i \frac{2\pi}{\lambda d} (x\xi + y\eta)\right] dx\, dy.$$ (1.25)

Since the object described by the function $b_1(x, y)$, has bounded dimensions, say $(-X_{max}, X_{max}; -Y_{max}, Y_{max})$ the function

$$\Gamma(\xi, \eta) \exp\left\{-i \frac{\pi}{\lambda d} [(\xi)^2 + (\eta)^2]\right\}$$

can be reconstructed through an interpolation of its readings:

$$\Gamma(\xi, \eta) \exp\left[-i \frac{\pi}{\lambda d} (\xi^2 + \eta^2)\right] = \sum_r \sum_s \Gamma(r\Delta\xi, s\Delta\eta) \times$$
$$\times \exp\left\{-i \frac{\pi}{\lambda d} [(r\Delta\xi)^2 + (s\Delta\eta)^2]\right\} \times$$
$$\times \mathrm{sinc}\left[\frac{\pi}{\Delta\xi} (\xi - r\Delta\xi)\right] \mathrm{sinc}\left[\frac{\pi}{\Delta\eta} (\eta - s\Delta\eta)\right],$$ (1.26)

where

$$\Delta\xi = \lambda d/2X_{max}, \quad \Delta\eta = \lambda d/2Y_{max}.$$ (1.27)

---

†Because of the discrete specification of $b_1(x, y)$ on the rectangular raster.

This means that $\Gamma(\xi, \eta)$ can also be reconstructed from its readings,

$$\Gamma(\xi, \eta) = \exp\left[i\frac{\pi}{\lambda d}(\xi^2 + \eta^2)\right]\sum_r\sum_s \Gamma(r\Delta\xi, s\Delta\eta) \times$$
$$\times \exp\left\{-i\frac{\pi}{\lambda d}[(r\Delta\xi)^2 + (s\Delta\eta)^2]\right\} \times$$
$$\times \text{sinc}\left[\frac{\pi}{\Delta\xi}(\xi - r\Delta\xi)\right]\text{sinc}\left[\frac{\pi}{\Delta\eta}(\eta - s\Delta\eta)\right] \qquad (1.28)$$

by means of an ordinary interpolating filter with phase correction on the plane on which this function is specified. This correction can be made, for example, by illuminating the hologram

$$\Gamma'(\xi, \eta) = \sum_r\sum_s \Gamma(r\Delta\xi, s\Delta\eta) \times$$
$$\times \exp\left\{-i\frac{\pi}{\lambda d}[(r\Delta\xi)^2 + (s\Delta\eta)^2]\right\} \times$$
$$\times \text{sinc}\frac{\pi}{\Delta\xi}(\xi - r\Delta\xi)\,\text{sinc}\frac{\pi}{\Delta\eta}(\eta - s\Delta\eta) \qquad (1.29)$$

with a spherical wavefront of the appropriate curvature,

$$\Gamma''(\xi, \eta) = \exp\left[i\frac{\pi}{\lambda d}(\xi^2 + \eta^2)\right]. \qquad (1.30)$$

Let us consider the procedure for finding the readings $\Gamma(r\Delta\xi, s\Delta\eta)$. From (1.25) we find

$$\Gamma(r\Delta\xi, s\Delta\eta)\exp\left\{-i\frac{\pi}{\lambda d}[(r\Delta\xi)^2 + (s\Delta\eta)^2]\right\} =$$
$$= \iint\limits_{(x,y)} b_1(x, y)\exp\left[i\frac{\pi}{\lambda d}(x^2 + y^2)\right] \times$$
$$\times \exp\left[-i\frac{2\pi}{\lambda d}(xr\Delta\xi + ys\Delta\eta)\right]dx\,dy. \qquad (1.31)$$

In the problem of synthesizing holograms, it is natural to assume that there is a smooth function $|b_1(x, y)|$, which describes the intensity reflectance of the object. Let us assume that $|b_1(x, y)|$ can be reconstructed from its readings through an interpolation based on some function $\phi(x, y)$:

$$|b_1(x, y)| = \sum_k\sum_l |b_1(k\Delta x, l\Delta y)|\,\phi(x - k\Delta x, y - l\Delta y), \qquad (1.32)$$

where Δx and Δy are the digitization intervals along the Cartesian coordinates x and y. Then

$$\Gamma\left(r\Delta\xi,\, s\Delta\eta\right) \exp\left\{-\,i\,\frac{\pi}{\lambda\,d}\,[(r\Delta\xi)^2 + (s\Delta\eta)^2]\right\} =$$
$$= \sum_k \sum_l |b_1(k\Delta x,\, l\Delta y)| \iint\limits_{(x,\,y)} \varphi(x - k\Delta x,\, y - l\Delta y).\times$$
$$\times \exp\left\{i\,\frac{\pi}{\lambda\,d}\,[\beta_1(x,\,y) + (x^2 + y^2)]\right\} \times$$
$$\times \exp\left\{-\,\frac{2\pi}{\lambda\,d}\,[xr\Delta\xi + ys\Delta\eta]\right\} dx\,dy, \tag{1.33}$$

where $\beta_1(x,y)$ is a function which is proportional to the phase of the reflection coefficient of the object, referred to a plane tangent to the object:

$$\beta_1(x,\,y) = \frac{\lambda d}{\pi}\,\beta(x,\,y). \tag{1.34}$$

It is not difficult to see that $\beta_1(x,y)$ describes the profile of the object with respect to this plane (see Fig. 1.3, which shows a cross section of the object and the tangent plane).

Within the sum and integral in (1.33) we introduce digitized phase factors which cancel each other out:

$$\Gamma\left(r\Delta\xi,\, s\Delta\eta\right) \exp\left\{-\,i\,\frac{\pi}{\lambda d}\,[(r\Delta\xi)^2 + (s\Delta\eta)^2]\right\} =$$
$$= \sum_k \sum_l |b_1(k\Delta x,\, l\Delta y)| \times$$
$$\times \exp\left\{i\,\frac{\pi}{\lambda d}\,[\beta_1(k\Delta x,\, l\Delta y) + (k\Delta x)^2 + (l\Delta y)^2]\right\} \times$$
$$\times \int\limits_{-X_{max}}^{X_{max}} \int\limits_{-Y_{max}}^{Y_{max}} \varphi(x - k\Delta x,\, y - l\Delta y) \times$$
$$\times \exp\left\{i\,\frac{\pi}{\lambda d}\,[\beta_1(x,\,y) - \beta_1(k\Delta x,\, l\Delta y)]\right\} \times$$
$$\times \exp\left\{i\,\frac{\pi}{\lambda d}\,[x^2 - (k\Delta x)^2 + y^2 - (l\Delta y)^2]\right\} \times$$
$$\times \exp\left[-\,i\,\frac{2\pi}{\lambda d}\,(xr\Delta\xi + ys\Delta\eta)\right] dx\,dy. \tag{1.35}$$

We now write the conditions

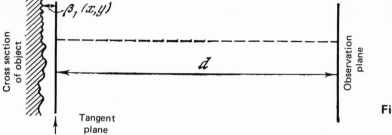

Fig. 1.3

$$\beta_1\,(x,\ y) - \beta_1\,(k\Delta x,\ l\Delta y) \ll \lambda d,$$
$$[x^2 - (k\Delta x)^2] + [y^2 - (l\Delta y)^2] \ll \lambda d. \qquad (1.36)$$

These conditions reduce to the condition that the error in conveying of the profile of the object and the spherical profile of the wavefront must be so small that we can write

$$\int\limits_{-X_{max}}^{X_{max}} \int\limits_{-Y_{max}}^{Y_{max}} \varphi\,(x - k\Delta x,\ y - l\Delta y) \times$$
$$\times \exp\left\{i\,\frac{\pi}{\lambda d}\,[\beta_1\,(x,\ y) - \beta_1\,(k\Delta x,\ l\Delta y)]\right\} \times$$
$$\times \exp\left\{i\,\frac{\pi}{\lambda d}\,[x^2 - (k\Delta x)^2 + y^2 - (l\Delta y)^2]\right\} \times$$
$$\times \exp\left[-\,i\,\frac{2\pi}{\lambda d}\,(xr\Delta\xi + ys\Delta\eta)\right] dx\,dy \cong$$
$$\cong \int\limits_{-X_{max}}^{X_{max}} \int\limits_{-Y_{max}}^{Y_{max}} \varphi\,(x - k\Delta x,\ y - l\Delta y) \times$$
$$\times \exp\left[-\,i\,\frac{2\pi}{\lambda d}\,(xr\Delta\xi + ys\Delta\eta)\right] dx\,dy. \qquad (1.37)$$

For given Δx and Δy, conditions (1.36) determine the minimum distance at which the object can be reconstructed correctly.

We transform the integral on the right side of (1.37) as follows:

$$\int\limits_{-X_{max}}^{X_{max}} \int\limits_{-Y_{max}}^{Y_{max}} \varphi\,(x - k\Delta x,\ y - l\Delta y) \times$$
$$\times \exp\left[-\,i\,\frac{2\pi}{\lambda d}\,(xr\Delta\xi + ys\Delta\eta)\right] dx\,dy =$$
$$= \left\{ \int\limits_{-X_{max}-k\Delta x}^{X_{max}-k\Delta x} \int\limits_{-Y_{max}-l\Delta y}^{Y_{max}-l\Delta y} \varphi\,(x,\ y) \times \right.$$

$$\times \exp\left[-i\,\frac{2\pi}{\lambda d}\,(xr\Delta\xi + ys\Delta\eta)\right]dx\,dy\right\} \times$$
$$\times \exp\left[-i\,\frac{2\pi}{\lambda d}\,(kr\Delta\xi\Delta x + ls\Delta\eta\Delta y)\right]. \qquad (1.38)$$

The dimension of the interpolation function $\varphi(x,y)$ along x and y is usually much smaller than the dimensions of the object of the interpolation ($2X_{max}$; $2Y_{max}$). The integral in (1.38) can thus be replaced by an integral with infinite limits, within an error associated with edge effects, which are significant only if $k\Delta x \cong \pm X_{max}, l\Delta y \cong \pm Y_{max}$. This replacement is equivalent to the application of a Fourier transformation of the interpolation function $\varphi(x,y)$:

$$\int_{-X_{max}-k\Delta x}^{X_{max}-k\Delta x}\int_{-Y_{max}-l\Delta y}^{Y_{max}-l\Delta y} \varphi(x,y) \times$$

$$\times \exp\left[-i\,\frac{2\pi}{\lambda d}\,(xr\Delta\xi + ys\Delta\eta)\right]dx\,dy \simeq \iint_{-\infty}^{\infty}\varphi(x,y) \times$$

$$\times \exp\left[-i\,\frac{2\pi}{\lambda d}\,(xr\Delta\xi + ys\Delta\eta)\right]dx\,dy = \Phi(r\Delta\xi, s\Delta\eta). \qquad (1.39)$$

Substituting (1.38) and (1.39) into (1.35), we find

$$\Gamma(r\Delta\xi, s\Delta\eta)\exp\left\{-i\,\frac{\pi}{\lambda d}\,[(r\Delta\xi)^2 + (s\Delta\eta)^2]\right\} =$$
$$= \left\{\sum_k\sum_l |b_1(k\Delta x, l\Delta y)| \times\right.$$

$$\times \exp\left\{i\,\frac{\pi}{\lambda d}\,[\beta_1(k\Delta x, l\Delta y) + (k\Delta x)^2 + (l\Delta y)^2]\right\} \times$$

$$\times \exp\left[-i\,\frac{2\pi}{\lambda d}\,(kr\Delta\xi\Delta x + ls\Delta\eta\Delta y)\right]\right\}\Phi(r\Delta\xi, s\Delta\eta). \qquad (1.40)$$

The summation over k and l in (1.40) is carried out over the ranges $[-X_{max}/\Delta x, X_{max}/\Delta x]$ and $[-Y_{max}/\Delta y, Y_{max}/\Delta y]$, respectively.

The function $(r\Delta\xi, s\Delta\eta)$ is the mask function: Outside a certain interval along ξ and η it vanishes. In an interpolation of the object on the basis of the sampling theorem we have

$$\varphi(x,y) = \text{sinc}\left(\frac{\pi}{\Delta x}\,x\right)\text{sinc}\left(\frac{\pi}{\Delta y}\,y\right), \qquad (1.41a)$$

so that

$$\Phi(r\Delta\xi, s\Delta\eta) = \begin{cases} 1; & |\xi| < \lambda d/\Delta x, \ |\eta| < \lambda d/\Delta y, \\ 0 & \text{otherwise}. \end{cases} \qquad (1.41b)$$

It follows that the maximum values of r and s for which the sum in (1.40) must be calculated are determined by

$$N_x = \lambda d / \Delta\xi \Delta x, \; N_y = \lambda d / \Delta\eta \Delta y. \tag{1.42}$$

The same quantities determine the number of terms in the sum in (1.40), so that, in accordance with (1.27),

$$2X_{max} / \Delta x = \lambda d / \Delta\xi \; \Delta x; \; 2Y_{max} / \Delta y = \lambda d / \Delta\eta \Delta y. \tag{1.43}$$

We thus have

$$\Gamma(r, s) = \exp\left\{ i \frac{\pi}{\lambda d} [r^2 (\Delta\xi)^2 + s^2 (\Delta\eta)^2] \right\} \times$$

$$\times \sum_{k=0}^{N_x-1} \sum_{l=0}^{N_y-1} b_1(k, l) \; \exp\left\{ i \frac{\pi}{\lambda d} [k^2 (\Delta x)^2 + l^2 (\Delta y)^2] \right\} \times$$

$$\times \exp\left[-i2\pi \left(\frac{kr}{N_x} + \frac{ls}{N_y} \right) \right]. \tag{1.44}$$

To eliminate the dimensionless quantities from (1.44), we introduce

$$\eta_\xi^2 = (2X_{max})^2 / \lambda d = \lambda d / (\Delta\xi)^2,$$

$$\eta_\eta^2 = (2Y_{max})^2 / \lambda d = \lambda d / (\Delta\eta)^2. \tag{1.45}$$

By virtue of (1.27), (1.42), and (1.43), we have

$$\frac{(\Delta x)^2}{\lambda d} = \frac{1}{N_x^2} \frac{(2X_{max})^2}{\lambda d} = \frac{\eta_\xi^2}{N_x^2},$$

$$\frac{(\Delta y)^2}{\lambda d} = \frac{1}{N_y^2} \frac{(2Y_{max})^2}{\lambda d} = \frac{\eta_\eta^2}{N_y^2}. \tag{1.46}$$

As a result we find

$$\Gamma(r, s) = \exp\left[i\pi \left(\frac{r^2}{\eta_\xi^2} + \frac{s^2}{\eta_\eta^2} \right) \right] \sum_{k=0}^{N_x-1} \sum_{l=0}^{N_y-1} b_1(k, l) \times$$

$$\times \exp\left[i\pi \left(\frac{k^2\eta_\xi^2}{N_x^2} + \frac{l^2\eta_\eta^2}{N_y^2} \right) \right] \exp\left[-i2\pi \left(\frac{kr}{N_x} + \frac{ls}{N_y} \right) \right]. \tag{1.47}$$

This equation is a discrete representation of a Fresnel transformation, reduced to a discrete Fourier transformation. In order to calculate it, we need the matrix $b_1(k, l)$, which determines the complex field amplitude at the object, and we also need to specify the quantities η_ξ and η_η, which are measures of the relative dimensions of the object as observed from the imaginary point at which the hologram is recorded. In this choice we should be guided by conditions (1.36) under which the error in the digitization of the phase factors is small.

The problem of synthesizing Fresnel holograms thus reduces to a calculation of the matrix $\{\Gamma(r, s)\}$ from the matrix of readings of the object, $\{b_1(k, l)\}$ and an analog interpolation of the readings obtained in accordance with (1.28)–(1.30). The interpolation in the reconstruction of the object from the hologram is carried out by masking the hologram with the function $(r \Delta\xi, s \Delta\eta)$ [see (1.39)].

Section 2. Theory of Fast Algorithms for Discrete Fourier Transformations

Fast Fourier Transformation. In this section we will consider methods for calculating DFTs,

$$\alpha_s = \frac{1}{\sqrt{N}} \sum_{k=0}^{N-1} a_k \exp\left(i2\pi \frac{ks}{N}\right), \qquad (2.1)$$

of the sequence $\{a_k\}$, where $k = 0, 1, \ldots, N - 1$. If the calculations are carried out on the basis of Eq. (2.1) directly, then to find all N coefficients α_s we must perform about N^2 complex operations $[N^2$ multiplications and $N(N - 1)$ complex additions]. Even for modest values of N this number can be very large, so that the DFT was useless for digital signal processing before the invention of the so-called fast Fourier transformation algorithms (FFTs).

The possibility of reducing the number of operations in a DFT calculation becomes obvious when we consider a two-dimensional DFT. For a two-dimensional array of dimensionality NM, the fact that a two-dimensional DFT can be factored into two one-dimensional DFTs means that the number of operations is approximately equal to N^2M for the DFT along the rows plus NM^2 for the DFT along the columns. In other words, there are $MN(M + N)$ operations, rather than $M^2N^2 = MN \times MN$, as there would be for a one-dimensional array of the same length. For an n-dimensional array of dimensionality $N_1 N_2 \ldots N_n$ the number of required operations is on the order of $N_1 N_2 \ldots N_n(N_1 + N_2 + \cdots + N_n)$, i.e., smaller by a factor of $(N_1 N_2 \ldots N_n)/(N_1 + \cdots + N_n)$ than the number corresponding to a one-dimensional array of the same volume. If we write the one-dimensional DFT as a multidimensional DFT, the number of operations is reduced markedly.

A one-dimensional transformation can be represented as a multidimensional transformation if, for example, the argument—the reading numbers—is treated as a vector. One possibility is to use a compound base for the number system for the reading numbers if the dimensionality of the array is a compound number, i.e., if N is the product of a certain number of simple factors:

$$N = N_1 N_2 \ldots N_n. \tag{2.2}$$

Let us discuss this possibility for the case of two factors:

$$N = N_1 N_2. \tag{2.3}$$

In this case, k and s can be written in a number system with bases N_1 and N_2:

$$k = k_2 N_1 + k_1, \quad s = s_1 N_2 + s_2, \tag{2.4}$$

where

$$k_1,\ s_1 = 0,1,\ldots,\ N_1 - 1;\ k_2,\ s_2 = 0,1,\ldots,\ N_2 - 1. \tag{2.5}$$

In this representation, the order of the digits of k and s is inverted:

$$k = (k_2, k_1),\ s = (s_1, s_2). \tag{2.6}$$

This is done deliberately, so that the bases for k and s will not be the same. Let us assume, for example, $N = 6$, so that we have $N_1 = 3$ and $N_2 = 2$. Then k and s in this number system, $(3, 2)$, are written as shown in Table 2.1.

Substitution of (2.4) into (2.1) yields

$$\alpha(s_1, s_2) = \frac{1}{\sqrt{N_1 N_2}} \sum_{k_2=0}^{N_2-1} \sum_{k_1=0}^{N_1-1} a(k_2, k_1) \exp\left[i2\pi \frac{(k_2 N_1 + k_1)(s_1 N_2 + s_2)}{N_2 N_1}\right] =$$

$$= \frac{1}{\sqrt{N_1 N_2}} \sum_{k_1=0}^{N_1-1} \exp\left[i2\pi\left(\frac{k_1 s_1}{N_1} + \frac{k_1 s_2}{N_1 N_2}\right)\right] \sum_{k_2=0}^{N_2-1} a(k_2, k_1) \exp\left[i2\pi \frac{k_2 s_2}{N_2}\right] =$$

$$= \frac{1}{\sqrt{N_1}} \sum_{k_1=0}^{N_1-1} A(k_1, s_2) \exp\left[i2\pi\left(\frac{k_1 s_1}{N_1} + \frac{k_1 s_2}{N_1 N_2}\right)\right] =$$

$$= \frac{1}{\sqrt{N_1}} \sum_{k_1=0}^{N_1-1} A_s(k_1, s_2) \exp\left(i2\pi \frac{k_1 s_1}{N_1}\right) = \mathrm{DFT}_{k_1}\{A_s(k_1, s_2)\} =$$

TABLE 2.1

$(k, s)_{10}$	0	1	2	3	4	5
k_2, k_1	00	01	02	10	11	12
s_1, s_2	00	01	10	11	20	21

$$= \text{DFT}_{k_1} \left\{ \exp\left(i2\pi \frac{k_1 s_2}{N} \right) \text{DFT}_{k_2} (a\,(k_2,\, k_1)) \right\}, \qquad (2.7)$$

where

$$A_s\,(k_1,\, s_2) = A\,(k_1,\, s_2)\, \exp\left(i2\pi \frac{k_1 s_2}{N} \right) =$$

$$= \frac{1}{\sqrt{N_2}} \exp\left(i2\pi \frac{k_1 s_2}{N} \right) \sum_{k_2=0}^{N_2-1} a\,(k_2,\, k_1)\, \exp\left(i2\pi \frac{k_2 s_2}{N_2} \right) =$$

$$= \exp\left(i2\pi \frac{k_1 s_2}{N} \right) \text{DFT}_{k_2} \{ a\,(k_2,\, k_1) \}. \qquad (2.8)$$

The original DFT is thus reduced to two DFTs, which are performed on the re-duced arrays. With this arrangement of calculations, the results of the calcula-tions—the coefficients α_s—are obtained in the "inverse" order, governed by the inversion of the digits in the position–number representation of their numbers in accordance with the compound base in (2.4) (see also Table 2.1).

Let us find the number of operations required for this calculation method. The discrete Fourier transformation $\text{DFT}_{k_2} \{ a\,(k_2,\, k_1) \}$ for one value of k_1 requires N_2^2 complex multiplications and $N_2(N_2 - 1)$ complex additions, and for all k_1 the corresponding number of complex multiplications is $N_2^2 N_1 = N_2 N$ and the corresponding number of complex additions is $N_2 N_1 (N_2 - 1) = N(N_2 - 1)$. Analogously, the transformation $\text{DFT}_{k_1} \{ A_s\,(k_1,\, s_2) \}$ re-quires N_1^2 complex multiplications and $N_1(N_1 - 1)$ complex additions for a single value of s_2; for all s_2, the required numbers of multiplications and addi-tions are $N_1^2 N_2 = N_1 N$ and $N_1(N_1 - 1)N_2 = (N_1 - 1)N$, respectively. The total number of multiplications is $N(N_1 + N_2)$ and the total number of additions is $N(N_1 + N_2 - 2)$, instead of $N^2 = NN_1 N_2$ and $N(N - 1) = N(N_1 N_2 - 1)$, respec-tively. If N_2 is also a compound number, then $\text{DFT}_{k_2} \{ a\,(k_1,\, k_2) \}$ can ob-viously be factored further. It can thus be concluded that in the general case

$$N = N_1 N_2 \ldots N_n, \qquad (2.9)$$

this procedure makes it possible to reduce the total number of operations in the

DFT calculation to a value on the order of $N \sum_{i=1}^{n} N_i$ instead of $N^2 = N \prod_{i=1}^{n} N_i$. It can thus be seen that the advantage of factoring becomes greater as the number of factors containing N increases. The maximum advantage is thus attained at that value of N which is comprised of the maximum number of minimum simple factors, i.e., twos:

$$N = 2^n. \tag{2.10}$$

In this case the number of required operations reduces from N^2 to $2Nn = 2N \log_2 N$. For $N = 2^{10}$, for example, the number of operations reduces from $2^{20} \approx 10^6$ to 10^4, i.e., by a factor of 100. The advantage is particularly apparent in the processing of two-dimensional arrays. For an array of 1000×1000 elements, a calculation of the spectrum without the use of FFTs would require about $2 \cdot 10^9$ complex operations; with a computer speed of, for example 10^4 complex operations per second, this processing would require about 55 h of computer time. When FFTs are used, the required computer time can be reduced to 0.55 h.

Graphical Representation of a Fast Fourier Transform. For clarity (and convenience in programming) a flow diagram can be drawn to correspond to the FFT algorithm. The elements of the diagram are the original readings of the signal and the results of the calculations. The lines with arrowheads connect the quantities to be summed and the result of the summation; the numbers beside the lines give the coefficient (unity is not shown).

The procedure for factoring a DFT which was described in the preceding subsection for the case $N = 6$ ($N_1 = 3, N_2 = 2$) is represented in Fig. 2.1. Also shown in this figure are the values of k_2, k_1, s_2, s_1. The coefficients of the summed readings are calculated in the first and second steps of the transformation as $(w^{N_1})^{k_2 s_1}$ and $(w^{N_2})^{k_1 s_1} w^{k_1 s_2}$, respectively, where $w = \exp\{i2\pi/6\}$.

This diagram quickly explains the physical meaning of the factoring: The original sequence is broken up into three subsequences, each consisting of two terms. A Fourier transformation is performed on each of the subsequences, and then the results are combined. This principle can be seen particularly clearly in the case $N = 2^n$ (see Fig. 2.2, where the analogous algorithm is shown as a diagram for $N = 2^3$). This method of breaking up the original sequence into subsequences has been called a "time decimation" [15] (a thinning out). It is also possible to invert the diagram and construct an FFT with frequency decimation. At present we have a variety of FFT algorithms which differ in the order of the input and output readings, in the way the memory is used (for example, the result can be found from the diagrams in Figs. 2.1 and 2.2 at the position of the input array; in other cases, an additional memory is required for intermediate arrays), and in other features. Various FFT versions can be found by interchanging the elements of the transformation diagram in any step. For example, an

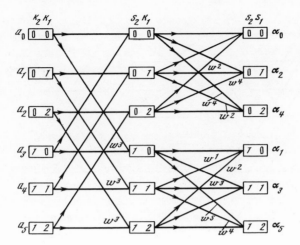

Fig. 2.1

interchange in the last step can eliminate the inversion in the readings of the result of the transformation (cf. Figs. 2.2 and 2.3).

Kronecker Matrices and Their Factorization. There is another way to approach the derivation of FFT algorithms—by using a matrix representation of the Fourier transformation. In this approach, the transformation matrix is factored into the product of several matrices which have a large number of zeros, and we get a better understanding of the orthogonal transformations which make "fast" algorithms possible.

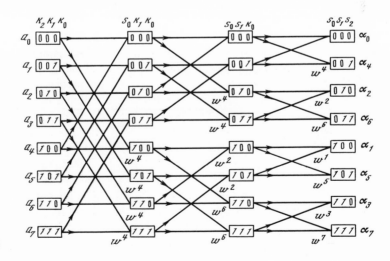

Fig. 2.2

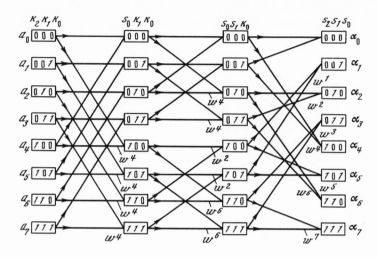

Fig. 2.3

We begin with the method for factoring matrices which are written as "direct" or "Kronecker" products of matrices of lower order [16].

The "direct product" of n matrices $M^{(i)}$ of order $m \times m$,

$$M^{[n]} = M^{(1)} \times M^{(2)} \times M^{(3)} \times \ldots \times M^{(n)} , \qquad (2.11)$$

is that matrix M of order $m^n \times m^n$, whose elements are defined by

$$\mu_{k, l}^{[n]} = \prod_{i=1}^{n} \mu_{k_i, l_i}^{(i)}, \qquad (2.12)$$

where k_i and l_i are the values of the digits in the position–number representation of the indices k and l with respect to base m:

$$k = \sum_{i=1}^{n} k_i m^{i-1}; \qquad l = \sum_{i=1}^{n} l_i m^{i-1}. \qquad (2.13)$$

For example,

$$M^{[2]} = \begin{pmatrix} A & B \\ C & D \end{pmatrix} \times \begin{pmatrix} a & b \\ c & d \end{pmatrix} = \begin{pmatrix} A \begin{pmatrix} a & b \\ c & d \end{pmatrix} & B \begin{pmatrix} a & b \\ c & d \end{pmatrix} \\ C \begin{pmatrix} a & b \\ c & d \end{pmatrix} & D \begin{pmatrix} a & b \\ c & d \end{pmatrix} \end{pmatrix} = \begin{pmatrix} Aa & Ab & Ba & Bb \\ Ac & Ad & Bc & Bd \\ Ca & Cb & Da & Db \\ Cc & Cd & Dc & Dd \end{pmatrix}, \quad (2.14)$$

In other words, the direct product of two matrices consists of submatrices which are equal to the product of the elements of the matrix at the left and of the matrix at the right. The matrices $M^{(i)}$ may be called the "generating matrices" for $M^{[n]}$; those matrices which are found by taking the direct product of n generating matrices are called the "Kronecker matrices of seniority n." According to the definition above, the generating matrices $M^{(i)}$ are multiplied from right to left. A change in their order in the direct product leads to an interchange of the digits in the position–number representation of the indices of the rows and columns of the resulting matrix and to a corresponding shift of the elements of this matrix.

An analogous definition can be offered for the direct product of matrices of different order (in this case the indices k and l of the elements of the resulting matrix are numbered in accordance with a compound base) and for the direct product of matrices which are not square.

Using the Kronecker delta, we can alter (2.12) by inserting all of the elements of all of the generating matrices and singling out the desired elements by means of the delta function. For generating matrices of different orders we thus find

$$\mu_{k,l}^{[n]} = \prod_{i=1}^{n} \prod_{r_i=0}^{K_i-1} \prod_{s_i=0}^{L_i-1} (\mu_{r_i s_i}^{(i)})^{\delta(k_i-r_i)\delta(l_i-s_i)}, \qquad (2.15)$$

where K_i, L_i is the order of the ith generating matrix; $\mu_{r_i s_i}$ is an element of the ith generating matrix; $k_i, r_i = 0, 1, \ldots, K_i - 1$; and $l_i, s_i = 0, 1, \ldots, L_i - 1$.

In the particular case in which all the generating matrices are square matrices of second rank,

$$M^{(i)} = \begin{pmatrix} A_i & B_i \\ C_i & D_i \end{pmatrix}, \qquad (2.16)$$

the delta functions in the exponent in (2.15) can be replaced by Boolean (logic) operations on the two-digit variables, k_i, r_i, l_i, s_i: $\delta(k_i - 0) = \delta(k_i) = \bar{k}_i$, $\delta(k_i - 1) = \delta(\bar{k}_i) = k_i$, etc., where the superior bar denotes a binary complementary minor (if $k_i = 0$, then $\bar{k}_i = 1$; if $k_i = 1$, then $\bar{k}_i = 0$). Equation (2.15) can be rewritten

$$\mu_{k,l}^{[n]} = \prod_{i=1}^{n} A_i^{\bar{k}_i \odot \bar{l}_i} B_i^{k_i \odot \bar{l}_i} C_i^{\bar{k}_i \odot l_i} D_i^{k_i \odot l_i}, \qquad (2.17)$$

and for identical 2 X 2 matrices we can write

$$\mu_{k,l}^{[n]} = A^{\sum\limits_{i=1}^{n} \bar{k}_i \odot \bar{l}_i} B^{\sum\limits_{i=1}^{n} k_i \odot \bar{l}_i} C^{\sum\limits_{i=1}^{n} \bar{k}_i \odot l_i} D^{\sum\limits_{i=1}^{n} k_i \odot l_i}, \qquad (2.18)$$

where $\odot$ represents a logic multiplication.

Here we have the following theorem [15, 16].

Theorem 1. The Kronecker matrix of seniority n which is found from the square $m \times m$ generating matrices,

$$M^{[n]} = M^{(1)} \times M^{(2)} \times \ldots \times M^{(n)}, \quad i = 1, 2, \ldots \quad n, \quad (2.19)$$

can be written as the product of n constituent matrices of order m^n,

$$M^{[n]} = M_c^{(1)} M_c^{(2)} \ldots M_c^{(n)}, \quad (2.20)$$

each of which is found from the corresponding generating matrix in accordance with the following rule: The sth row of the generating matrix is written in the $(sm^{n-1} + k)$th row of the constituent matrix, beginning from the element of this matrix with index km^{n-1}, where $k = 0, 1, \ldots,$ with $m - 1; s = 0, 1, \ldots, m - 1;$ the other elements of the given row of the constituent matrix are assumed to be zero:

$$M_c^{(i)} = \begin{pmatrix}
\mu_{0,0}^{(i)} & \mu_{0,1}^{(i)} & \cdots & \mu_{0,m-1}^{(i)} & 0 & 0 & . & . & . & . & 0 \\
0 & 0 & \cdots & 0 & \mu_{0,0}^{(i)} & \mu_{0,1}^{(i)} & \cdots & \mu_{0,m-1}^{(i)} & 0 & \cdots & 0 \\
. & . & . & . & . & . & . & . & . & . & . \\
0 & 0 & . & . & . & . & 0 & \mu_{0,0}^{(i)} & \cdots & \mu_{0,m-1}^{(i)} \\
\mu_{1,0}^{(i)} & \mu_{1,1}^{(i)} & \cdots & \mu_{1,m-1}^{(i)} & 0 & 0 & . & . & . & . & 0 \\
0 & 0 & \cdots & 0 & \mu_{1,0}^{(i)} & \mu_{1,1}^{(i)} & \cdots & \mu_{1,m-1}^{(i)} & 0 & \cdots & 0 \\
. & . & . & . & . & . & . & . & . & . & . \\
0 & 0 & . & . & . & . & 0 & \mu_{1,0}^{(i)} & \cdots & \mu_{1,m-1}^{(i)} \\
. & . & . & . & . & . & . & . & . & . & . \\
\mu_{m-1,0}^{(i)} & \mu_{m-1,1}^{(i)} & \cdots & \mu_{m-1,m-1}^{(i)} & 0 & 0 & . & . & . & . & 0 \\
0 & 0 & \cdots & 0 & \mu_{m-1,0}^{(i)} & \mu_{m-1,1}^{(i)} & \cdots & \mu_{m-1,m-1}^{(i)} & 0 & \cdots & 0 \\
. & . & . & . & . & . & . & . & . & . & . \\
0 & 0 & . & . & . & . & 0 & \mu_{m-1,0}^{(i)} & \cdots & \mu_{m-1,m-1}^{(i)}
\end{pmatrix} \quad (2.21)$$

The elements of the ith constituent matrix can be written in general as

$$\mu_c^{(i)}(k, l) = \mu_{k_{n-1}, l_0}^{(i)} \delta\left(\sum_{s=1}^{n-1} k_{s-1} m^s - \sum_{s=1}^{n-1} l_s m^s\right) =$$

$$= \mu_{k_{n-1}, l_0}^{(i)} \delta(k_{n-2} - l_{n-1}) \delta(k_{n-3} - l_{n-2}) \ldots \delta(k_0 - l_1), \quad (2.22)$$

where k_s and l_s are digits of the indices k and l in their position–number representation in terms of the base m:

$$k = \sum_{s=0}^{n-1} k_s m^s, \quad l = \sum_{s=0}^{n-1} l_s m^s. \quad (2.23)$$

The theorem can be proven by mathematical induction.

Example. We assume

$$M^{[2]} = \begin{pmatrix} A & B \\ C & D \end{pmatrix} \times \begin{pmatrix} a & b \\ c & d \end{pmatrix} = \begin{pmatrix} Aa & Ab & Ba & Bb \\ Ac & Ad & Bc & Bd \\ Ca & Cb & Da & Db \\ Cc & Cd & Dc & Dd \end{pmatrix}. \qquad (2.24)$$

We check the product of the corresponding constituent matrices:

$$\begin{pmatrix} AB & 0 \\ 0 & AB \\ CD & 0 \\ 0 & CD \end{pmatrix} \begin{pmatrix} ab & 0 \\ 0 & ab \\ cd & 0 \\ 0 & cd \end{pmatrix} = \begin{pmatrix} Aa & Ab & Ba & Bb \\ Ac & Ad & Bc & Bd \\ Ca & Cb & Da & Db \\ Cc & Cd & Dc & Dd \end{pmatrix}. \qquad (2.25)$$

This factorization of the Kronecker matrices has the obvious advantage that the constituent matrices of order $m^n \times m^n$ contain the largest of the m^{n+1} nonzero elements. Accordingly, in order to multiply them n times by a column vector of order m^n, it is necessary to carry out only nm^{n+1} operations instead of the m^{2n} operations required in an immediate multiplication of the vector by the matrix $m^{[n]}$. This factorization gives us an algorithm for a rapid multiplication of Kronecker matrices.

There is an analogous theorem for the Kronecker product of matrices of different order. In this case the constituent matrices have orders $m_1 \times m_2 \times \cdots \times m_n$, where m_i is the order of the ith generating matrix. The ith constituent matrix is constructed from the ith generating matrix in the following manner: The columns of the constituent matrix are broken up into $(\prod_{s=1}^{n} m_s)/m_i$ groups of columns, each containing m_i columns, while the rows are broken up into m_i groups of rows, each containing $(\prod_{s=1}^{n} m_s)/m_i$ rows. As a result, this matrix is written as $(\prod_{s=1}^{n} m_s)/m_i$ submatrices of order m_i. This approach corresponds to the following indexing of the rows and columns of the constituent matrix:

$$k = k_1 \left[\left(\prod_{s=1}^{n} m_s \right) \bigg/ m_i \right] + k_0, \quad l = l_0 m_i + l_1, \qquad (2.26)$$

where $k_0, l_0 = 0, 1, \ldots, [(\prod_{s=1}^{n} m_s)/m_i] - 1$; and $k_1 l_1 = 0, 1, \ldots m_i - 1$. We now assume that $\mu_{k_1 l_1}^{(i)}$ is an element with index k_1, l_1 of the ith generating matrix. Then the element of the ith constituent matrix with index (k, l) is

$$\mu_c^{(i)} (k, l) = \mu_{k_1 l_1}^{(i)} \delta (k_0 - l_0). \qquad (2.27)$$

Direct Matrix Product of the Second Kind and Factorization of Kronecker Matrices. In order to analyze DFTs, the concept of the direct (Kronecker) matrix product can be supplemented with another concept, which is a generalization of the direct product in a certain sense. In the case of the direct product, the generating matrix which is second in order is multiplied in a scalar manner by the elements of the first matrix. We use the term "direct product of the second kind" or "matrix direct product" of matrices of order $r \times p$ and $np \times mr$ $(n = 1, 2, \ldots; m = 1, 2, \ldots)$ to refer to that matrix M of order $np \times mr$,

$$M = M_{r,p} \otimes M_{np,mr}, \tag{2.28}$$

which is found by breaking up the matrix $M_{np,mr}$ into submatrices $M_{p,r}^{i,j}$ of order $p \times r$ and multiplying $M_{r,p}$ by these submatrices:

$$
\begin{aligned}
M &= M_{r,p} \otimes M_{np,mr} = \\
&= M_{r,p} \otimes \begin{pmatrix} M_{p,r}^{0,0} & M_{p,r}^{(0,1)} \cdots \\ M_{p,r}^{(1,0)} & M_{p,r}^{(1,1)} \cdots \\ \cdots \cdots \cdots & M_{p,r}^{(n-1,m-1)} \end{pmatrix} = \\
&= \begin{pmatrix} (M_{r,p} \cdot M_{p,r}^{(0,0)}) (M_{r,p} \cdot M_{p,r}^{(0,1)}) & \cdots \cdots \cdots \\ (M_{r,p} \cdot M_{p,r}^{(1,0)}) (M_{r,p} \cdot M_{p,r}^{(1,1)}) & \cdots \cdots \cdots \\ \cdots \cdots \cdots \cdots \cdots \cdots \cdots \cdots \cdots \cdots \\ \cdots \cdots \cdots \cdots \cdots \cdots M_{r,p} \cdot M_{p,r}^{(n-1,m-1)} \end{pmatrix}. \tag{2.29}
\end{aligned}
$$

The direct product of the second kind of the matrix $M_{nr,mp} \otimes M_{p,r}$ can be defined in an analogous manner.†

Now we can state a theorem which relates the direct product of the second kind to the ordinary matrix product and to the direct product of the first kind.

Theorem 2. Let M_r and M_{nr} be square matrices of order $r \times r$ and $nr \times nr$, respectively. Then

$$
\begin{aligned}
M_r \otimes M_{nr} &= (I_n \times M_r)M_{nr}, \\
M_{nr} \otimes M_r &= M_{nr}(I_n \times M_r), \tag{2.30}
\end{aligned}
$$

where I_n is the $n \times n$ matrix

$$I_n = \{\delta(i-j)\} = \begin{pmatrix} 1 & 0 & 0 \ldots 0 \\ 0 & 1 & 0 \ldots 0 \\ 0 & 0 & 1 \ldots 0 \\ \cdot & \cdot & \cdot \cdot \cdot \cdot \\ 0 & & \cdot \cdot \cdot \cdot \cdot 1 \end{pmatrix}, \quad i, j = 0, 1, \ldots, n-1. \tag{2.31}$$

Theorem 3. Let $M_{r_1}^{(1)}$ and $M_{r_2}^{(2)}$ be square matrices of order $r_1 \times r_1$ and $r_2 \times r_2$, respectively. Then

---

†In the case $n = 1, m = 1$ the direct product of the second kind is the same as the ordinary matrix product.

$$M_{r_1}^{(1)} \times M_{r_2}^{(2)} = (M_{r_1}^{(1)} \times I_{r_2}) \otimes M_{r_2}^{(2)}. \tag{2.32}$$

The proof of these theorems is almost obvious: The direct product $I_n \times M_r$ in (2.30) gives a matrix with submatrices along the diagonal and zeros elsewhere, so that in multiplication by M_{nr} we obtain the product of M_r and the submatrices M_{nr}. In (2.32), the direct product $M_{r_1}^{(1)} \times I_{r_2}$ gives us a matrix with diagonal submatrices, and the multiplication of these submatrices by $M_{r_2}^{(2)}$ is equivalent to the multiplication of $M_{r_2}^{(2)}$ by the elements of the matrix $M_{r_1}^{(1)}$.

A corollary of Theorems 2 and 3 is another theorem, which describes yet another method for factoring Kronecker matrices [16].

Theorem 4. Let $M_{r_0}^0$, $M_{r_1}^{(1)}$, ..., $M_{r_{k-1}}^{(k-1)}$ by the square matrices $r_0 \times r_0$, $r_1 \times r_1, \ldots, r_{k-1} \times r_{k-1}$. Then

$$\begin{aligned} M_{r_0}^{(0)} \times M_{r_1}^{(1)} \times \ldots \times M_{r_i}^{(i)} \times \ldots \times M_{r_{k-1}}^{(k-1)} = (M_{r_0}^{(0)} \times I_{r_1} \times \\ \times I_{r_2} \times \ldots \times I_{r_{k-1}})(I_{r_0} \times M_{r_1}^{(1)} \times I_{r_2} \times \ldots \times I_{r_{k-1}}) \ldots \\ \ldots (I_{r_0} \times I_{r_1} \times \ldots \times I_{r_{i-1}} \times M_{r_i}^{(i)} \times I_{r_{i+1}} \times \ldots \times I_{r_{k-2}}) \ldots \\ \ldots (I_{r_0} \times I_{r_1} \times \ldots \times I_{r_{k-2}} \times M_{r_{k-1}}^{(k-1)}). \end{aligned} \tag{2.33}$$

To prove this theorem we can use the following property of the direct product:

$$\begin{aligned} (M^{(1)} \times M^{(2)} \times \ldots \times M^{(k)})(N^{(1)} \times N^{(2)} \times \ldots \times N^{(k)}) \ldots \\ \ldots (P^{(1)} \times P^{(2)} \times \ldots \times P^{(k)}) = (M^{(1)}N^{(1)} \ldots P^{(1)}) \times \\ \times (M^{(2)}N^{(2)} \ldots P^{(2)}) \times \ldots \times (M^{(k)}N^{(k)} \ldots P^{(k)}). \end{aligned} \tag{2.34}$$

This property is easily proved by mathematical induction:

$$\begin{aligned} (M_{r_0}^{(0)} \times I_{r_1} \times \ldots \times I_{r_k}) \ldots (I_{r_0} \times \ldots \times I_{r_{k-1}} \times M_{r_k}^{(k)}) = \\ = (M_{r_0}^{(0)} I_{r_0} \ldots I_{r_0}) \times (I_{r_1} M_{r_1} I_{r_1} \ldots I_{r_1}) \times \ldots \\ \ldots \times (I_{r_k} \ldots I_{r_k} M_{r_k}^{(k)}) = M_{r_0}^{(0)} \times M_{r_1}^{(1)} \times \ldots \times M_{r_k}^{(k)}. \end{aligned}$$

The Discrete Fourier Transformation and FFT Algorithms from the Standpoint of the Factorization of Kronecker Matrices. There is a profound relationship between the matrix of the DFT and the Kronecker matrices. To see this relationship we consider the example of the DFT matrix F_8:

$$F_8 = \left\{\exp\left(i2\pi \frac{rs}{8}\right)\right\} = \{w^{rs}\} = \begin{vmatrix} w^0 & w^0 & w^0 & w^0 & w^0 & w^0 & w^0 & w^0 \\ w^0 & w^1 & w^2 & w^3 & w^4 & w^5 & w^6 & w^7 \\ w^0 & w^2 & w^4 & w^6 & w^8 & w^{10} & w^{12} & w^{14} \\ w^0 & w^3 & w^6 & w^9 & w^{12} & w^{15} & w^{18} & w^{21} \\ w^0 & w^4 & w^8 & w^{12} & w^{16} & w^{20} & w^{24} & w^{28} \\ w^0 & w^5 & w^{10} & w^{15} & w^{20} & w^{25} & w^{30} & w^{35} \\ w^0 & w^6 & w^{12} & w^{18} & w^{24} & w^{30} & w^{36} & w^{42} \\ w_0 & w^7 & w^{14} & w^{21} & w^{28} & w^{35} & w^{42} & w^{49} \end{vmatrix} . \tag{2.35}$$

We perform a binary inversion of its rows (this inversion corresponds to a binary inversion of the Fourier coefficients found as the result of the transformation†), and we simplify the matrix elements by making use of

$$w^k = w^{(k) \bmod 8}; \; w^0 = 1. \tag{2.36}$$

Then,

$$F_{8\,\text{inv}} = \begin{pmatrix} 1 & 1 & 1 & 1 & 1 & 1 & 1 & 1 \\ 1 & w^4 & 1 & w^4 & 1 & w^4 & 1 & w^4 \\ 1 & w^2 & w^4 & w^6 & 1 & w^2 & w^4 & w^6 \\ 1 & w^6 & w^4 & w^2 & 1 & w^6 & w^4 & w^2 \\ 1 & w^1 & w^2 & w^3 & w^4 & w^5 & w^6 & w^7 \\ 1 & w^5 & w^2 & w^7 & w^4 & w^1 & w^6 & w^3 \\ 1 & w^3 & w^6 & w^1 & w^4 & w^7 & w^2 & w^5 \\ 1 & w^7 & w^6 & w^5 & w^4 & w^3 & w^2 & w^1 \end{pmatrix}. \tag{2.37}$$

We now break up $F_{8\,inv}$ into four submatrices and transform them, again making use of (2.36):

$$F_{8\,\text{inv}} = \left(\begin{array}{cccc:cccc} 1 & 1 & 1 & 1 & 1 & 1 & 1 & 1 \\ 1 & w^4 & 1 & w^4 & 1 & w^4 & 1 & w^4 \\ 1 & w^2 & w^4 & w^6 & 1 & w^2 & w^4 & w^6 \\ 1 & w^6 & w^4 & w^2 & 1 & w^6 & w^4 & w^2 \\ \hdashline 1 & w^1 & w^2 & w^3 & w^4 & w^5 & w^6 & w^7 \\ 1 & w^5 & w^2 & w^7 & w^4 & w^1 & w^6 & w^3 \\ 1 & w^3 & w^6 & w^1 & w^4 & w^7 & w^2 & w^5 \\ 1 & w^7 & w^6 & w^5 & w^4 & w^3 & w^2 & w^1 \end{array} \right) =$$

$$= \left(\begin{array}{cccc:cccc} 1 & 1 & 1 & 1 & 1 & 1 & 1 & 1 \\ 1 & w^4 & 1 & w^4 & 1 & w^4 & 1 & w^4 \\ 1 & w^2 & w^4 & w^6 & 1 & w^2 & w^4 & w^6 \\ 1 & w^6 & w^4 & w^2 & 1 & w^6 & w^4 & w^2 \\ \hdashline 1 & w^1 & w^2 & w^3 & w^4 & w^5 & w^6 & w^7 \\ 1 & w^5 & w^2 & w^7 & w^4 & w^9 & w^6 & w^{11} \\ 1 & w^3 & w^6 & w^9 & w^4 & w^7 & w^{10} & w^{13} \\ 1 & w^7 & w^6 & w^5 & w^4 & w^{11} & w^{10} & w^9 \end{array} \right). \tag{2.38}$$

It is not difficult to see that the two upper submatrices are the same, while the

---

†We recall that for F_8 in a binary row inversion, the zeroth (000), second (010), fifth (101), and seventh (111) rows retain their places, while the first (001) and fourth (100) rows are interchanged, as are the third (011) and sixth (110).

lower two are found from them by multiplying the columns by the coefficients $1, w^1, w^2, w^3, w^4, w^5, w^6, w^7$, respectively. Since the multiplication of the columns of a matrix by certain numbers can be represented as the product of this matrix and a diagonal matrix with these numbers on the diagonal, we can write

$$
F_{8\,\mathrm{inv}} = \left(
\begin{array}{c}
\left(\left(\begin{pmatrix} 1 & 1 & 1 & 1 \\ 1 & w^4 & 1 & w^4 \\ 1 & w^2 & w^4 & w^6 \\ 1 & w^6 & w^4 & w^2 \end{pmatrix} \cdot \begin{pmatrix} 1 & 0 & 0 & 0 \\ 0 & 1 & 0 & 0 \\ 0 & 0 & 1 & 0 \\ 0 & 0 & 0 & 1 \end{pmatrix}\right) \vdots \left(\begin{pmatrix} 1 & 1 & 1 & 1 \\ 1 & w^4 & 1 & w^4 \\ 1 & w^2 & w^4 & w^6 \\ 1 & w^6 & w^2 & w^4 \end{pmatrix} \cdot \begin{pmatrix} 1 & 0 & 0 & 0 \\ 0 & 1 & 0 & 0 \\ 0 & 0 & 1 & 0 \\ 0 & 0 & 0 & 1 \end{pmatrix}\right)\right) \\[2em]
\left(\left(\begin{pmatrix} 1 & 1 & 1 & 1 \\ 1 & w^4 & 1 & w^4 \\ 1 & w^2 & w^4 & w^6 \\ 1 & w^6 & w^4 & w^2 \end{pmatrix} \cdot \begin{pmatrix} 1 & 0 & 0 & 0 \\ 0 & 1 & 0 & 0 \\ 0 & 0 & 1 & 0 \\ 0 & 0 & 0 & 1 \end{pmatrix}\right) \vdots \left(\begin{pmatrix} 1 & 1 & 1 & 1 \\ 1 & w^4 & 1 & w^4 \\ 1 & w^2 & w^4 & w^6 \\ 1 & w^6 & w^2 & w^4 \end{pmatrix} \cdot \begin{pmatrix} w^4 & 0 & 0 & 0 \\ 0 & w^5 & 0 & 0 \\ 0 & 0 & w^6 & 0 \\ 0 & 0 & 0 & w^7 \end{pmatrix}\right)\right)
\end{array}
\right) =
$$

$$
= \begin{pmatrix} 1 & 1 & 1 & 1 \\ 1 & w^4 & 1 & w^4 \\ 1 & w^2 & w^4 & w^6 \\ 1 & w^6 & w^4 & w^2 \end{pmatrix} \otimes \begin{pmatrix} 1 & & & & 1 & & & \\ & 1 & & & & 1 & & \\ & & 1 & & & & 1 & \\ & & & 1 & & & & 1 \\ 1 & & & & w^4 & & & \\ & w^1 & & & & w^5 & & \\ & & w^2 & & & & w^6 & \\ & & & w^3 & & & & w^7 \end{pmatrix} =
$$

$$
= \begin{pmatrix} 1 & 1 & 1 & 1 \\ 1 & w^4 & 1 & w^4 \\ 1 & w^2 & w^4 & w^6 \\ 1 & w^6 & w^4 & w^2 \end{pmatrix} \otimes \left[\begin{pmatrix} 1 & & & & \\ & 1 & & & \\ & & 1 & & \\ & & & w^1 & \\ & & & & w^2 \\ & & & & & w^3 \end{pmatrix} \cdot \begin{pmatrix} 1 & & 1 & & \\ & 1 & & 1 & \\ & & 1 & & 1 \\ 1 & & w^4 & & \\ & 1 & & w^4 & \\ & & 1 & & w^4 \end{pmatrix}\right]
$$

$$
= \begin{pmatrix} 1 & 1 & 1 & 1 \\ 1 & w^4 & 1 & w^4 \\ 1 & w^2 & w^4 & w^6 \\ 1 & w^6 & w^4 & w^2 \end{pmatrix} \otimes \left[\begin{pmatrix} 1 & & & \\ & 1 & & \\ & & 1 & \\ & & & w^1 \\ & & & & w^2 \\ & & & & & w^3 \end{pmatrix} \left[\begin{pmatrix} 1 & 1 \\ 1 & w^4 \end{pmatrix} \times \begin{pmatrix} 1 & 0 \\ 0 & 1 \end{pmatrix} \times \begin{pmatrix} 1 & 0 \\ 0 & 1 \end{pmatrix}\right]\right].
$$

(2.39)

We now replace the direct product of the second kind by the Kronecker product and a simple matrix product:

$$
F_{8\,\mathrm{inv}} = \left[\begin{pmatrix} 1 & 0 \\ 0 & 1 \end{pmatrix} \times \begin{pmatrix} 1 & 1 & 1 & 1 \\ 1 & w^4 & 1 & w^4 \\ 1 & w^2 & w^4 & w^6 \\ 1 & w^6 & w^4 & w^2 \end{pmatrix}\right].
$$

$$\cdot \begin{pmatrix} 1 & & & & & & & \\ & 1 & & & & & & \\ & & 1 & & & & & \\ & & & 1 & & & & \\ & & & & 1 & & & \\ & & & & & w^1 & & \\ & & & & & & w^2 & \\ & & & & & & & w^3 \end{pmatrix} \left[\begin{pmatrix} 1 & 1 \\ 1 & w^4 \end{pmatrix} \times \begin{pmatrix} 1 & 0 \\ 0 & 1 \end{pmatrix} \times \begin{pmatrix} 1 & 0 \\ 0 & 1 \end{pmatrix} \right]. \tag{2.40}$$

In precisely the same way, we can factor the matrix

$$\begin{pmatrix} 1 & 1 & 1 & 1 \\ 1 & w^4 & 1 & w^4 \\ 1 & w^2 & w^4 & w^6 \\ 1 & w^6 & w^4 & w^2 \end{pmatrix}, \tag{2.41}$$

breaking it up into submatrices and writing it in the form of a direct product of the second kind:

$$F_{8\,\text{inv}} = \left\{ \begin{pmatrix} 1 & 0 \\ 0 & 1 \end{pmatrix} \times \left[\begin{pmatrix} 1 & 1 \\ 1 & w^4 \end{pmatrix} \otimes \left[\begin{pmatrix} 1 & & \\ & 1 & \\ & & w^2 \end{pmatrix} \left[\begin{pmatrix} 1 & 1 \\ 1 & w^4 \end{pmatrix} \times \begin{pmatrix} 1 & 0 \\ 0 & 1 \end{pmatrix} \right] \right] \right] \right\} \cdot$$

$$\cdot \begin{pmatrix} 1 & & & & & & & \\ & 1 & & & & & & \\ & & 1 & & & & & \\ & & & 1 & & & & \\ & & & & 1 & & & \\ & & & & & w^1 & & \\ & & & & & & w^2 & \\ & & & & & & & w^3 \end{pmatrix} \left[\begin{pmatrix} 1 & 1 \\ 1 & w^4 \end{pmatrix} \times \begin{pmatrix} 1 & 0 \\ 0 & 1 \end{pmatrix} \times \begin{pmatrix} 1 & 0 \\ 0 & 1 \end{pmatrix} \right] =$$

$$= \left\{ \begin{pmatrix} 1 & 0 \\ 0 & 1 \end{pmatrix} \times \left[\left[\begin{pmatrix} 1 & 0 \\ 0 & 1 \end{pmatrix} \times \begin{pmatrix} 1 & 1 \\ 1 & w^4 \end{pmatrix} \right] \begin{pmatrix} 1 & & \\ & 1 & \\ & & w^2 \end{pmatrix} \cdot \left[\begin{pmatrix} 1 & 1 \\ 1 & w^4 \end{pmatrix} \times \begin{pmatrix} 1 & 0 \\ 0 & 1 \end{pmatrix} \right] \right] \right\} \cdot$$

$$\cdot \begin{pmatrix} 1 & & & & & & & \\ & 1 & & & & & & \\ & & 1 & & & & & \\ & & & 1 & & & & \\ & & & & 1 & & & \\ & & & & & w^1 & & \\ & & & & & & w^2 & \\ & & & & & & & w^3 \end{pmatrix} \left[\begin{pmatrix} 1 & 1 \\ 1 & w^4 \end{pmatrix} \times \begin{pmatrix} 1 & 0 \\ 0 & 1 \end{pmatrix} \times \begin{pmatrix} 1 & 0 \\ 0 & 1 \end{pmatrix} \right]. \tag{2.42}$$

Finally, using (2.34) and the identity $I_n = I_n I_n \ldots$ to bring the unit matrix at the left into the matrix multiplication, we find a completely factored representation for the matrix $F_{8\,inv}$:

$$
F_{8\,inv} = \left[\begin{pmatrix} 1 & 0 \\ 0 & 1 \end{pmatrix} \times \begin{pmatrix} 1 & 0 \\ 0 & 1 \end{pmatrix} \times \begin{pmatrix} 1 & 1 \\ 1 & w^4 \end{pmatrix} \right] \left[\begin{pmatrix} 1 & 0 \\ 0 & 1 \end{pmatrix} \times \begin{pmatrix} 1 & & \\ & 1 & \\ & & 1 \\ & & & w^2 \end{pmatrix} \right] \cdot
$$

$$
\cdot \left[\begin{pmatrix} 1 & 0 \\ 0 & 1 \end{pmatrix} \times \begin{pmatrix} 1 & 1 \\ 1 & w^4 \end{pmatrix} \times \begin{pmatrix} 1 & 0 \\ 0 & 1 \end{pmatrix} \right] \begin{pmatrix} 1 \\ & 1 \\ & & 1 \\ & & & 1 \\ & & & & 1 \\ & & & & & w^1 \\ & & & & & & w^2 \\ & & & & & & & w^3 \end{pmatrix} \cdot
$$

$$
\cdot \left[\begin{pmatrix} 1 & 1 \\ 1 & w^4 \end{pmatrix} \times \begin{pmatrix} 1 & 0 \\ 0 & 1 \end{pmatrix} \times \begin{pmatrix} 1 & 0 \\ 0 & 1 \end{pmatrix} \right]. \tag{2.43}
$$

Now, noting that $w^4 = \exp(i\pi) = -1$, we can write

$$
F_{8\,inv} = \left(\begin{pmatrix} 1 & 0 \\ 0 & 1 \end{pmatrix} \times \begin{pmatrix} 1 & 0 \\ 0 & 1 \end{pmatrix} \times \begin{pmatrix} 1 & 1 \\ 0 & -1 \end{pmatrix} \right) \cdot \left(\begin{pmatrix} 1 & 0 \\ 0 & 1 \end{pmatrix} \times \begin{pmatrix} 1 & & \\ & 1 & \\ & & 1 \\ & & & w^2 \end{pmatrix} \right) \cdot
$$

$$
\cdot \left(\begin{pmatrix} 1 & 0 \\ 0 & 1 \end{pmatrix} \times \begin{pmatrix} 1 & 1 \\ 1 & -1 \end{pmatrix} \times \begin{pmatrix} 1 & 0 \\ 0 & 1 \end{pmatrix} \right) \cdot \begin{pmatrix} 1 \\ & 1 \\ & & 1 \\ & & & 1 \\ & & & & 1 \\ & & & & & w^1 \\ & & & & & & w^2 \\ & & & & & & & w^3 \end{pmatrix} \cdot
$$

$$
\cdot \left(\begin{pmatrix} 1 & 1 \\ 1 & -1 \end{pmatrix} \times \begin{pmatrix} 1 & 0 \\ 0 & 1 \end{pmatrix} \times \begin{pmatrix} 1 & 0 \\ 0 & 1 \end{pmatrix} \right). \tag{2.44}
$$

Figure 2.4 is a diagram of the transformation corresponding to the factorization in (2.44). The structure of this diagram is the same as that in Fig. 2.2, but the coefficients along the lines are different. This is, therefore, a different modification of the algorithm. By using it and performing some simple operations on the matrices, we can construct several other modifications. These operations are quite instructive. We can illustrate them for the case of the transformation to the diagram shown in Fig. 2.2.

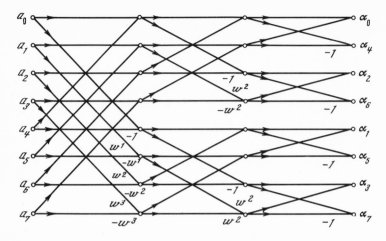

Fig. 2.4

To keep the transformations from becoming too complicated, we denote the constituent matrices in (2.44) as $F_8^{(0)}, F_8^{(1)}, F_8^{(2)}, F_3^{(3)}, F_8^{(4)}$. Then

$$F_{8\,\text{inv}} = F_8^{(0)} F_8^{(1)} F_8^{(2)} F_8^{(3)} F_8^{(4)}. \tag{2.45}$$

We assume that M_8^{in} is a matrix which performs a binary inversion of rows. Each row of this matrix has a one in the column whose index is the binary-inverted index of the row. The other elements of this row are zero:

$$M_8^{\text{in}} = \begin{pmatrix} 1 & 0 & 0 & 0 & 0 & 0 & 0 & 0 \\ 0 & 0 & 0 & 0 & 1 & 0 & 0 & 0 \\ 0 & 0 & 1 & 0 & 0 & 0 & 0 & 0 \\ 0 & 0 & 0 & 0 & 0 & 0 & 1 & 0 \\ 0 & 1 & 0 & 0 & 0 & 0 & 0 & 0 \\ 0 & 0 & 0 & 0 & 0 & 1 & 0 & 0 \\ 0 & 0 & 0 & 1 & 0 & 0 & 0 & 0 \\ 0 & 0 & 0 & 0 & 0 & 0 & 0 & 1 \end{pmatrix}. \tag{2.46}$$

We note that

$$M_8^{\text{in}} M_8^{\text{in}} = I_8, \tag{2.47}$$

and, in general,

$$(M_8^{\text{in}})^{2n} = I_8. \tag{2.48}$$

Then

$$F_8 = M_8^{\text{in}} F_8^{(0)} F_8^{(1)} F_8^{(2)} F_8^{(3)} F_8^{(4)}. \qquad (2.49)$$

Let us transpose F_8. Since the DFT matrix is symmetric with respect to its main diagonal,

$$F_8^T = F_8 = (F_8^{(4)})^T (F_8^{(3)})^T (F_8^{(2)})^T (F_8^{(1)})^T (F_8^{(0)})^T (M_8^{\text{in}})^T. \qquad (2.50)$$

Furthermore, since all the constituent matrices are symmetric,

$$F_8 = F_8^{(4)} F_8^{(3)} F_8^{(2)} F_8^{(1)} F_8^{(0)} M_8^{\text{in}}. \qquad (2.51)$$

This is yet another modification of the algorithm. It corresponds to the diagram which would be found from Fig. 2.2 by rotating from right to left. This algorithm obviously provides for a binary inversion of the input sequence. Also,

$$
\begin{aligned}
F_{8\,\text{inv}} &= M_8^{\text{in}} F_8^{(4)} F_8^{(3)} F_8^{(2)} F_8^{(1)} F_8^{(0)} M_8^{\text{in}} = \\
&= (M_8^{\text{in}} F_8^{(4)} M_8^{\text{in}})(M_8^{\text{in}} F_8^{(3)} M_8^{\text{in}})(M_8^{\text{in}} F_8^{(2)} M_8^{\text{in}}) \cdot \\
&\quad \cdot (M_8^{\text{in}} F_8^{(1)} M_8^{\text{in}})(M_8^{\text{in}} F_8^{(0)} M_8^{\text{in}}).
\end{aligned} \qquad (2.52)
$$

We now note that a matrix of the type

$$M_a = M^{\text{in}} M_b M^{\text{in}} \qquad (2.53)$$

is the matrix M_b with a simultaneous inversion of rows and columns. If M_b is Kronecker matrix, then M_a is also a Kronecker matrix, with the generating matrices in the opposite order. We thus have

$$M_8^{\text{in}} F_8^{(4)} M_8^{\text{in}} = \begin{pmatrix} 1 & 0 \\ 0 & 1 \end{pmatrix} \times \begin{pmatrix} 1 & 0 \\ 0 & 1 \end{pmatrix} \times \begin{pmatrix} 1 & 1 \\ 1 & -1 \end{pmatrix} = F_8^{(0)}, \qquad (2.54)$$

$$M_8^{\text{in}} F_8^{(3)} M_8^{\text{in}} = \begin{pmatrix} 1 & & & & & & & \\ & 1 & & & & & & \\ & & 1 & & & & & \\ & & & w^2 & & & & \\ & & & & 1 & & & \\ & & & & & w^1 & & \\ & & & & & & 1 & \\ & & & & & & & w^3 \end{pmatrix}, \qquad (2.55)$$

$$M_8^{\text{in}} F_8^{(2)} M_8^{\text{in}} = \begin{pmatrix} 1 & 0 \\ 0 & 1 \end{pmatrix} \times \begin{pmatrix} 1 & 1 \\ 1 & -1 \end{pmatrix} \times \begin{pmatrix} 1 & 0 \\ 0 & 1 \end{pmatrix} = F_8^{(2)}, \qquad (2.56)$$

$$M_8^{\text{in}} F_8^{(1)} M_8^{\text{in}} = \begin{pmatrix} 1 & & & & & & & \\ & 1 & & & & & & \\ & & 1 & & & & & \\ & & & 1 & & & & \\ & & & & 1 & & & \\ & & & & & 1 & & \\ & & & & & & w^2 & \\ & & & & & & & w^2 \end{pmatrix}, \qquad (2.57)$$

$$M_8^{\text{in}} F_8^{(0)} M_8^{\text{in}} = \begin{pmatrix} 1 & 1 \\ 1 & -1 \end{pmatrix} \times \begin{pmatrix} 1 & 0 \\ 0 & 1 \end{pmatrix} \times \begin{pmatrix} 1 & 0 \\ 0 & 1 \end{pmatrix} = F_8^{(4)}, \qquad (2.58)$$

$$F_8^{\text{in}} = \begin{pmatrix} 1 & 1 & & & & & & \\ 1 & -1 & & & & & & \\ & & 1 & 1 & & & & \\ & & 1 & -1 & & & & \\ & & & & 1 & 1 & & \\ & & & & 1 & -1 & & \\ & & & & & & 1 & 1 \\ & & & & & & 1 & -1 \end{pmatrix} \begin{pmatrix} 1 & & & & & & & \\ & 1 & & & & & & \\ & & 1 & & & & & \\ & & & w^2 & & & & \\ & & & & 1 & & & \\ & & & & & w^1 & & \\ & & & & & & 1 & \\ & & & & & & & w^3 \end{pmatrix} \cdot$$

$$\cdot \begin{pmatrix} 1 & 0 & 1 & 0 & & & & \\ 0 & 1 & 0 & 1 & & & & \\ 1 & 0 & -1 & 0 & & & & \\ 0 & 1 & 0 & -1 & & & & \\ & & & & 1 & 0 & 1 & 0 \\ & & & & 0 & 1 & 0 & 1 \\ & & & & 1 & 0 & -1 & 0 \\ & & & & 0 & 1 & 0 & -1 \end{pmatrix} \cdot \begin{pmatrix} 1 & & & & & & & \\ & 1 & & & & & & \\ & & 1 & & & & & \\ & & & 1 & & & & \\ & & & & 1 & & & \\ & & & & & 1 & & \\ & & & & & & w^2 & \\ & & & & & & & w^2 \end{pmatrix} \cdot$$

$$\cdot \begin{pmatrix} 1 & & & & 1 & & & \\ & 1 & & & & 1 & & \\ & & 1 & & & & 1 & \\ & & & 1 & & & & 1 \\ 1 & & & & -1 & & & \\ & 1 & & & & -1 & & \\ & & 1 & & & & -1 & \\ & & & 1 & & & & -1 \end{pmatrix}. \qquad (2.59)$$

It is simple to see that this is also an algorithm which corresponds to the diagram in Fig. 2.2.

Returning to Eq. (2.44), we note that it can be shown that this equation is a particular case of the general equation

$$F_{2^n} = M_{2^n}^{in} \prod_{i=0}^{n-1} (I_{2^{n-i-1}} \times D_{2^{i+1}}^{(i)})(I_{2^{n-i-1}} \times h_2 \times I_{2^i}), \qquad (2.60)$$

where I_k is a $k \times k$ unit matrix, and $k \times k$; $D_{2^{i+1}}^{(i)}$ is a $2^{i+1} \times 2^{i+1}$ diagonal matrix,

$$D_{2^{i+1}}^{(i)} = \{d_{rs} = (w^{2^{n-i-1}})^{r_i \sum\limits_{-\infty}^{i-1} r_k 2^k} \delta(r-s)\}, \ r, s = 0, 1, \cdots \\ \cdots, 2^{i+1} - 1; \qquad (2.61)$$

where r_i and r_k are the binary digits in the binary representation of the row number r, and

$$h_2 = \begin{pmatrix} 1 & 1 \\ 1 & -1 \end{pmatrix} \qquad (2.62)$$

is the generating matrix of the Hadamard matrix. The cofactors are numbered from left to right. This expression is a compact form of one of the possible FFT algorithms, and it can also be used for a straightforward derivation of any other FFT algorithms.

To conclude this subsection we will describe a method for constructing an FFT modification in which a factorization of Kronecker matrices in accordance with Theorem 1 is used. As the initial representation we choose the transposed version of (2.60):

$$F_{2^n} = \left(\prod_{i=n-1}^{0} (I_{2^{n-i-1}} \times h_2 \times I_{2^i})(I_{2^{n-i-1}} \times D_{2^{i+1}}^{(i)}) \right) M_{2^n}^{in}. \qquad (2.63)$$

We transform the matrices involved in the direct products into constituent matrices, making use of the first theorem on the factorization of Kronecker matrices. To keep the equations compact, we denote these transformed matrices by the superscript "G":

$$F_{2^n} = \left(\prod_{i=n-1}^{0} (I_{2^{n-i-1}})_{2^n}^G (h_2)_{2^n}^G (I_{2^i})_{2^n}^G (I_{2^{n-i-1}} \times D_{2^{i+1}}^{(i)}) \right) M_{2^n}^{in}. \qquad (2.64)$$

We can regroup the factors to single out the constituent matrices constructed from h_2:

$$F_{2^n} = \left(\prod_{i=n-1}^{0} (h_2)_{2^n}^G [(I_{2^i})_{2^n}^G (I_{2^{n-i-1}} \times D_{2^{i+1}}^{(i)})(I_{2^{n-i}})_{2^n}^G] \right) M_{2^n}^{in}. \qquad (2.65)$$

It is not difficult to see that matrices of the type $(I_{2^i})_{2^n}^G$ contain precisely 2^n ones and $2^n(2^n - 1)$ zeros and are transposed matrices. The matrix of the product at the right in the brackets in (2.65) interchanges the columns of the internal matrix, while that at the left interchanges the rows. These two transposition matrices complement each other in a certain sense: The matrix $(I_{2^{n-i}})_{2^n}^G$ has $2^{n-i} \times 2^i$ submatrices consisting of $2^i \times 2^{n-i}$ elements, while the matrix $(I_{2^i})_{2^n}^G$ has $2^i \times 2^{n-i}$ submatrices consisting of $2^{n-i} \times 2^i$ elements, which are transposed with respect to the corresponding submatrices $(I_{2^{n-i}})_{2^n}^G$. As a result, the order of the interchange of rows and columns in the internal matrix is the same. Since this is a diagonal matrix, the transposition matrices leave it diagonal, but its elements are interchanged. We can find the rule for the transformation of the indices of the diagonal elements by recalling how the constituent matrix is constructed on the basis of Theorem 1. Let us consider the matrix at the left, $(I_{2^{i-1}})_{2^n}^G$. The index of each row of this matrix is the same as the index of a certain element on the diagonal of the internal matrix. This is the element whose position is occupied (as the result of the transposition) by that element whose index is the same as that of the element of this row which is equal to one. Let us assume that k is the index of the row of this matrix (and thus the index of the element on the transformed diagonal of the internal matrix:

$$k = \sum_{s=0}^{n-1} k_s 2^s. \tag{2.66}$$

We write k in the form

$$k = \sum_{s=n-i}^{n-1} k_s 2^s + \sum_{s=0}^{n-i-1} k_s 2^s = \left(\sum_{p=0}^{i-1} k_{n+p-i} 2^p \right) 2^{n-i} +$$
$$+ \sum_{s=0}^{n-i-1} k_s 2^s = x_1 2^{n-1} + x_0, \tag{2.67}$$

where

$$x_1 = \sum_{p=0}^{i-1} k_{n+p-i} 2^p, \quad x_0 = \sum_{s=0}^{n-i-1} k_s 2^s. \tag{2.68}$$

Now the index l of the element of this row which is equal to one can be written

$$l = \lambda_0 2^i + \lambda_1, \tag{2.69}$$

where

$$\lambda_0 = \sum_{p=0}^{n-i-1} l_{p+i} 2^p = x_0 = \sum_{s=0}^{n-i-1} k_s 2^s,$$

$$\lambda_1 = \sum_{s=0}^{i-1} l_s 2^s = x_1 = \sum_{p=0}^{i-1} k_{n+p-i} 2^p \qquad (2.70)$$

and $\{l_s\}$ are the values of the binary digits in the binary representation of l:

$$l = \sum_{s=0}^{n-1} l_s 2^s. \qquad (2.71)$$

We can now state a rule for the transformation of the binary digits $\{l_s\}$ of an element of the internal diagonal matrix in (2.65):

$$
\begin{aligned}
l_{r+i} &= k_r; & r &= 0, 1, \ldots, n - i - 1; \\
l_p &= k_{n+p-i}; & p &= 0, 1, \ldots, i + 1,
\end{aligned}
\qquad (2.72)
$$

in other words, the sequence $\{l_s\}$ is broken up into the two groups $\{l_{n-1}, l_{n-2}, \ldots, l_i\}$ and $\{l_{i-1}, \ldots, l_0\}$, which contain $n - i$ and i digits, respectively. These groups are interchanged, forming the new number $\{l_{i-1}, \ldots, l_0, l_{n-1}, \ldots, l_i\}$ which is the index of that element of the diagonal which is reached by the element with index $\{l_s\}$.

Let us illustrate this transformation for the case $n = 3$. With $i = 0$, the diagonal matrix does not change [in the case $i = 0$, there are ones on the diagonal according to (2.61) and (2.65)]. With $i = 1$, the transformation table is

000	001	010	011	100	101	110	111
↓	↓	↓	↓	↓	↓	↓	↓
000	100	001	101	010	110	011	111;

in the case $i = 2$, it is

000	001	010	011	100	101	110	111
↓	↓	↓	↓	↓	↓	↓	↓
000	010	100	110	001	011	101	111.

As a result, we find the following FFT algorithm from (2.65):

$$
F_8 = \begin{pmatrix}
1 & 1 & 0 & 0 & 0 & 0 & 0 & 0 \\
0 & 0 & 1 & 1 & 0 & 0 & 0 & 0 \\
0 & 0 & 0 & 0 & 1 & 1 & 0 & 0 \\
0 & 0 & 0 & 0 & 0 & 0 & 1 & 1 \\
1 & -1 & 0 & 0 & 0 & 0 & 0 & 0 \\
0 & 0 & 1 & -1 & 0 & 0 & 0 & 0 \\
0 & 0 & 0 & 0 & 1 & -1 & 0 & 0 \\
0 & 0 & 0 & 0 & 0 & 0 & 1 & -1
\end{pmatrix} \cdot \begin{pmatrix}
1 & & & & & & & \\
& 1 & & & & & & \\
& & 1 & & & & & \\
& & & w^1 & & & & \\
& & & & 1 & & & \\
& & & & & w^2 & & \\
& & & & & & 1 & \\
& & & & & & & w^3
\end{pmatrix} \cdot
$$

$$\cdot \begin{pmatrix} 1 & 1 & 0 & 0 & 0 & 0 & 0 & 0 \\ 0 & 0 & 1 & 1 & 0 & 0 & 0 & 0 \\ 0 & 0 & 0 & 0 & 1 & 1 & 0 & 0 \\ 0 & 0 & 0 & 0 & 0 & 0 & 1 & 1 \\ 1 & -1 & 0 & 0 & 0 & 0 & 0 & 0 \\ 0 & 0 & 1 & -1 & 0 & 0 & 0 & 0 \\ 0 & 0 & 0 & 0 & 1 & -1 & 0 & 0 \\ 0 & 0 & 0 & 0 & 0 & 0 & 1 & -1 \end{pmatrix} \cdot \begin{pmatrix} 1 & & & & & & & \\ & 1 & & & & & & \\ & & 1 & & & & & \\ & & & 1 & & & & \\ & & & & 1 & & & \\ & & & & & w^2 & & \\ & & & & & & 1 & \\ & & & & & & & w^2 \end{pmatrix} \cdot$$

$$\cdot \begin{pmatrix} 1 & 1 & 0 & 0 & 0 & 0 & 0 & 0 \\ 0 & 0 & 1 & 1 & 0 & 0 & 0 & 0 \\ 0 & 0 & 0 & 0 & 1 & 1 & 0 & 0 \\ 0 & 0 & 0 & 0 & 0 & 0 & 1 & 1 \\ 1 & -1 & 0 & 0 & 0 & 0 & 0 & 0 \\ 0 & 0 & 1 & -1 & 0 & 0 & 0 & 0 \\ 0 & 0 & 0 & 0 & 1 & -1 & 0 & 0 \\ 0 & 0 & 0 & 0 & 0 & 0 & 1 & -1 \end{pmatrix} \cdot M_8^{\text{in}}. \tag{2.73}$$

The transformation diagram corresponding to this algorithm is shown in Fig. 2.5. The primary advantage of this diagram is that the structure is identical for all steps.

Truncated FFT Algorithms. When a DFT is used, it frequently turns out that either the original sequence contains many zero elements or it is not necessary to calculate all the transformation coefficients (or both). These circumstances can be exploited to further reduce the number of operations required for the transformation. Examining the FFT diagrams, we see that either of these circum-

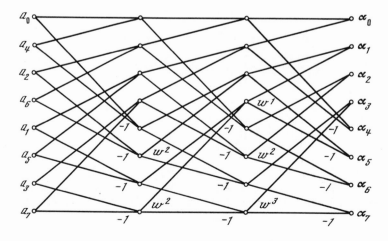

Fig. 2.5

stances means that certain lines of the transformation diagram drop out; i.e., some of the multiplication and addition operations drop out. If the number of zero elements in the transformed sequence or the number of unnecessary transformation elements is less than half the length of the sequence, there is no substantial effect on the structure of the transformation diagram, and the reduction in the number of operations is small. In the opposite case, the reduction in the number of operations can be appreciable.

Let us determine how the structure of the FFT algorithms changes in this case, and let us estimate the possible advantage. For definiteness, we adopt (2.63) as the original representation:

$$F_{2^n} = \left(\prod_{i=n-1}^{0} (I_{2^{n-i-1}} \times h_2 \times I_{2^i})(I_{2^{n-i-1}} \times D_{2^{i+1}}^{(i)}) \right) M_{2^n}^{\text{in}}. \qquad (2.74)$$

The presence of zero elements in the transformed sequence and of unnecessary transformation coefficients can be described conveniently by multiplying the transformation matrix from the right and from the left by diagonal "cutting" matrices (which we denote by $P_{2^n}^{(k)}$ and $P_{2^n}^{(l)}$, respectively), which contain zeros at those points on the diagonal which correspond to the indices of the zero elements or the unnecessary elements.

From the standpoint of the possibility of modifying the transformation algorithms, it is sufficient to consider the case in which the number of nonzero or necessary elements is an integral power of two. In this case the cutting matrices can be written as Kronecker products of matrices of the type

$$P_2 = \begin{pmatrix} 1 & 0 \\ 0 & 0 \end{pmatrix} \qquad (2.75)$$

and unit matrices,

$$I_2 = \begin{pmatrix} 1 & 0 \\ 0 & 1 \end{pmatrix}. \qquad (2.76)$$

We consider the simplest case, in which the last elements of the original sequence and their transformation are cut out. Then the cutting matrix of rank 2^n, which has 2^l ones at the beginning of its diagonal, is written

$$P_{2^n}^{(l)} = \prod_{n-l} \mathbf{d}\, P_2 \times \prod_{l} \mathbf{d}\, I_2, \qquad (2.77)$$

where $\prod \mathbf{d}$ denotes the direct matrix product, to be contrasted with $\prod$, the ordinary matrix product.

We seek a factored form of the matrix F_{2^n}, truncated on two sides, by using two approaches. We first find $P_{2^n}^{(k)} F_{2^n}$, and then $P_{2^n}^k F_{2^n} P_{2^n}^l$.

$$P_{2^n}^{(k)} F_{2^n} = \left(\prod_{n-k} \mathbf{d}\, P_2 \times \prod_k \mathbf{d}\, I_2 \right) \cdot$$

$$\cdot \left(\prod_{i=n-1}^{0} (I_{2^{n-i-1}} \times h_2 \times I_{2^i}) \cdot (I_{2^{n-i-1}} \times D_{2^i+1}^{(i)}) \right) M_{2^n}^{\text{in}}. \quad (2.78)$$

Using property (2.34) of the product of Kronecker matrices for the product of the first two matrices on the left, we find

$$\left(\prod_{n-k} \mathbf{d}\, P_2 \times \prod_k \mathbf{d}\, I_2 \right) (h_2 \times I_{2^{n-1}}) = (P_2 \cdot h_2) \times \prod_{n-k-1} \mathbf{d}\, P_2 \times$$

$$\times \prod_k \mathbf{d}\, I_2 = \begin{pmatrix} 1 & 1 \\ 0 & 0 \end{pmatrix} \times \prod_{n-k-1} \mathbf{d}\, P_2 \times \prod_k \mathbf{d}\, I_2 =$$

$$= \begin{pmatrix} 1 & 1 \\ 0 & 0 \end{pmatrix} \begin{pmatrix} 1 & 0 \\ 0 & 1 \end{pmatrix} \times \prod_{n-k-1} \mathbf{d}\, (P_2 P_2) \times \prod_k \mathbf{d}\, (I_2 I_2) =$$

$$= \left(S_2 \times \prod_{n-k-1} \mathbf{d}\, P_2 \times \prod_k \mathbf{d}\, I_2 \right) \left(I_2 \times \prod_{n-k-1} \mathbf{d}\, P_2 \times \prod_k \mathbf{d}\, I_2 \right), \quad (2.79)$$

where

$$S_2 = \begin{pmatrix} 1 & 1 \\ 0 & 0 \end{pmatrix}. \quad (2.80)$$

The matrix on the right in (2.79) is a diagonal cutting matrix. If we substitute (2.79) into (2.78), we find that this matrix is followed by the diagonal matrix Since the product of diagonal matrices is commutative, and the matrix

$$I_2 \times \prod_{n-k-1} \mathbf{d}\, P_2 \times \prod_k \mathbf{d}\, I_2$$

is a cutting matrix, it turns some of the elements in $D_{2^n}^{(n-1)}$, into zero and thus moves on to the next step ($i = n - 2$) of the product in (2.78). After arguments of this type, we easily find

$$P_{2^n}^{(k)} F_{2^n} = \left(\prod_{i=n-1}^{k} (I_{2^{n-i-1}} \times S_2 \times P_{2^i}^{(k)}) \cdot \right.$$

$$\cdot [(I_{2^{n-i}} \times P_{2^i}^{(k)})(I_{2^{n-i-1}} \times D_{2^i+1}^{(i)})] \right) \cdot$$

$$\cdot \left(\prod_{i=k-1}^{0} (I_{2^{n-i-1}} \times h_2 \times I_{2^i})(I_{2^{n-i-1}} \times D_{2^i+1}^{(i)}) \right) M_{2^n}^{\text{in}}. \quad (2.81)$$

We can now find a factored form of the FFT matrix for the case in which this matrix is also cut from the right. We first interchange the cutting matrix and the binary-inversion matrix, which is the last matrix in the product in (2.74):

$$P_{2^n}^{(k)} F_{2^n} P_{2^n}^{(l)} = P_{2^n}^{(k)} F_{2^n}^{(\text{in})} M_{2^n}^{\text{in}} P_{2^n}^{(l)} =$$

$$= P_{2^n}^{(k)} F_{2^n}^{\text{in}} M_{2^n}^{\text{in}} P_{2^n}^{(l)} (M_{2^n}^{\text{in}} M_{2^n}^{\text{in}}) =$$

$$= P_{2^n}^{(k)} F_{2^n}^{\text{in}} (M_{2^n}^{\text{in}} P_{2^n}^{(l)} M_{2^n}^{\text{in}}) M_{2^n}^{\text{in}} = P_{2^n}^{(k)} F_{2^n}^{\text{in}} \widetilde{P}_{2^n}^{(l)} M_{2^n}^{\text{in}}, \qquad (2.82)$$

where

$$\widetilde{P}_{2^n}^{(l)} = \prod_l \mathbf{d}\, I_2 \times \prod_{n-l} \mathbf{d}\, P_2 = I_{2^l} \times P_{2^{n-l}} \qquad (2.83)$$

is found from the matrix $P_{2^n}^{(l)}$ by changing the order of the generating matrices [the result of multiplying $P_{2^n}^{(l)}$ in (2.82) from the left and right by the binary-inversion matrix].

We turn now to the product $\widetilde{P}_{2^n}^{(l)}$ and to the first constituent matrix on the right in (2.81), $P_{2^n}^k F_{2^n}^{\text{in}}$:

$$(I_{2^{n-1}} \times h_2)(I_{2^{n-1}} \times D_2^{(0)})(I_{2^l} \times P_{2^{n-l}}) =$$

$$= (I_{2^{n-1}} \times h_2)(I_{2^l} \times P_{2^{n-l}})[(I_{2^{n-1}} \times D_2^{(0)})(I_{2^l} \times P_{2^{n-l}})] =$$

$$= (I_{2^l} \times P_{2^{n-l-1}} \times S_2^T)[(I_{2^{n-1}} \times D_2^{(0)})(I_{2^l} \times P_{2^{n-l}})] =$$

$$= (I_{2^l} \times P_{2^{n-l-1}} \times I_2)(I_{2^l} \times P_{2^{n-l-1}} \times S_2^T) \cdot$$

$$\cdot [(I_{2^{n-1}} \times D_2^{(0)})(I_{2^l} \times P_{2^{n-l}})], \qquad (2.84)$$

where

$$S_2^T = \begin{pmatrix} 1 & 0 \\ 1 & 0 \end{pmatrix}. \qquad (2.85)$$

The matrix $(I_{2^l} \times P_{2^{n-l-1}} \times I_2)$, in the product on the left in (2.84) is multiplied by the next constituent matrix $(i = 1)$ in (2.81), and so forth. For the case $k + l \geqslant n$ we have

$$P_{2^n}^{(k)} F_{2^n} P_{2^n}^{(l)} = \Big(\prod_{i=n-1}^{k} (I_{2^{n-i-1}} \times S_2 \times P_{2^{i-k}} \times I_{2^k}) \cdot$$

$$\cdot [(I_{2^{n-i}} \times P_{2^{i-k}} \times I_{2^k})(I_{2^{n-i-1}} \times D_{2^{i+1}}^{(i)})] \Big) \cdot$$

$$\cdot \prod_{i=k-1}^{n-l} (I_{2^{n-i-1}} \times h_2 \times I_{2^i})(I_{2^{n-i-1}} \times D_{2^{i+1}}^{(i)}) \cdot$$

$$\cdot \Big(\prod_{i=n-l-1}^{0} (I_{2^l} \times P_{2^{n-l-i-1}} \times S_2^T \times I_{2^i}) \cdot$$

$$\cdot [(I_{2^{n-i-1}} \times D_{2^{i+1}}^i)(I_{2^l} \times P_{2^{n-l-i}} \times I_{2^i})] \Big) M_{2^n}^{\text{in}} =$$

$$= \Big(\prod_{i=n-1}^{k} (I_{2^{n-i-1}} \times S_2 \times P_{2^i}^{(k)}) \times [(I_{2^{n-i-1}} \times I_2 \times P_{2^i}^{(k)})(I_{2^{n-i-1}} \times D_{2^{i+1}}^{(i)})] \Big) \cdot$$

$$\cdot \prod_{i=k-1}^{n-l} (I_{2^{n-i-1}} \times h_2 \times I_{2^i})(I_{2^{n-i-1}} \times D_{2^{i+1}}^{(i)}) \cdot$$

$$\cdot \Big(\prod_{i=n-l-1}^{0} (I_{2l} \times P_{2^{n-l-i-1}} \times S_2^T \times I_{2^i}) \cdot$$

$$\cdot [(I_{2^{n-i-1}} \times D_{2^{i+1}}^{(i)})(I_{2^l} \times P_{2^{n-l-i-1}} \times P_{2^{i+1}}^{(i)})] \Big) M_{2^n}^{\text{in}} =$$

$$= \Big(\prod_{i=n-1}^{k} (I_{2^{n-i-1}} \times S_2 \times P_{2^i}^{(k)}) [I_{2^{n-i-1}} \times ((I_2 \times P_{2^i}^{(k)}) D_{2^{i+1}}^{(i)})] \Big) \cdot$$

$$\cdot \prod_{i=k-1}^{n-l} (I_{2^{n-i-1}} \times h_2 \times I_{2^i})(I_{2^{n-i-1}} \times D_{2^{i+1}}^{(i)}) \cdot$$

$$\cdot \Big\{ \prod_{i=n-l-1}^{0} (I_{2^l} \times P_{2^{n-l-i-1}} \times S_2^T \times I_{2^i}) \cdot$$

$$\cdot [(I_{2^l} \times (P_{2^{n-l-i-1}} \times (D_{2^{i+1}}^{(i)} P_{2^{i+1}}^{(i)})))] \Big\} M_{2^n}^{\text{in}}. \tag{2.86}$$

Using Eq. (2.61) we can simplify this equation:

$$D_{2^{i+1}}^{(i)} = \{d_{r,\,s} = (w^{2^{n-i-1}})^{r_i \sum\limits_{0}^{i-1} r_j 2^j} \delta(r-s)\},$$

$$r, s = 0, 1, \ldots, 2^{i+1} - 1; \; r = \sum_{j=0}^{i} r_j \cdot 2^j. \tag{2.87}$$

Then

$$(I_2 \times P_{2^i}^{(k)}) D_{2^{i+1}}^{(i)} = (I_2 \times P_{2^i}^{(k)}) D_{2^{i+1}}^{(i,\,k)}, \tag{2.88}$$

where

$$D_{2^{i+1}}^{(i,\,k)} = \{d_{r,\,s}^{(i,\,k)} = (w^{2^{n-i-1}})^{r_i \sum\limits_{0}^{k-1} r_j 2^j} \delta(r-s)\} \tag{2.89}$$

and

$$P_{2^{n-l-i-1}} \times (D^{(i)}_{2^{i+1}} P^{(i)}_{2^{i+1}}) = P^{(i)}_{2^{n-l}}. \tag{2.90}$$

For $k + l \geqslant n$ we then find

$$P^{(k)}_{2^n} F_{2^n} P^{(l)}_{2^n} = \left[\left(\prod_{i=n-1}^{k} (I_{2^{n-i-1}} \times S_2 \times P^{(k)}_{2^i}) \cdot \right. \right.$$
$$\left. \cdot [I_{2^{n-i-1}} \times ((I_2 \times P^{(k)}_{2^i}) D^{(i,\,k)}_{2^{i+1}})] \right) \cdot$$
$$\cdot \prod_{i=k-1}^{n-l} (I_{2^{n-i-1}} \times h_2 \times I_{2^i})(I_{2^{n-i-1}} \times D^{(i)}_{2^{i+1}}) \cdot$$
$$\left. \cdot \prod_{i=n-l-1}^{0} (I_{2^l} \times P_{2^{n-l-i-1}} \times S_2^T \times I_{2^i})(I_{2^l} \times P^{(i)}_{2^{n-l}}) \right] M^{\text{in}}_{2^n}. \tag{2.91}$$

Finally, we recall that

$$\prod_{i=n-l-1}^{0} (I_{2^l} \times P_{2^{n-l-i-1}} \times S_2^T \times I_{2^i})(I_{2^l} \times P^{(i)}_{2^{n-l}}) =$$
$$= I_{2^l} \times \prod_{n-l} \mathrm{d}\, S_2^T = I_{2^l} \times S^T_{2^{n-l}}, \tag{2.92}$$

where $S^T_{2^{n-l}}$ is a matrix of rank 2^{n-l} whose first column consists of ones, while all the other columns consist of zeros. The matrices $I_{2^l} \times S^T_{2^{n-l}}$ are duplication matrices: It is easily shown that their multiplication of a vector leads to a $2^{(n-l)}$-fold duplication (repetition) of those vector components whose indices are multiples of $2^{(n-l)}$ and to the discarding of the other components. As a result we finally find, for $k + l \geqslant n$,

$$P^{(k)}_{2^n} F_{2^n} P^{(l)}_{2^n} = \left\{ \prod_{i=n-1}^{k} (I_{2^{n-i-1}} \times S_2 \times P^{(k)}_{2^i}) \cdot \right.$$
$$\left. \cdot [I_{2^{n-i-1}} \times ((I_2 \times P^{(k)}_{2^i}) D^{(i,\,k)}_{2^{i+1}})] \right\} \cdot$$
$$\cdot \left[\prod_{i=k-1}^{n-l} (I_{2^{n-i-1}} \times h_2 \times I_{2^i})(I_{2^{n-i-1}} \times D^{(i)}_{2^{i+1}}) \right] \cdot (I_{2^l} \times S^T_{2^{n-l}}) M^{\text{in}}_{2^n}. \tag{2.93}$$

With $k + l \geqslant n$, the truncated Fourier transformation is thus factored into the product of $l + 1$ matrices of three types (not counting the binary-inversion matrices): the duplication matrices, the $(k + l - n)$ ordinary constituent FFT matrices with indices i from $k - 1$ to $n - l$, and the $n - k$ truncated constituent FFT matrices with i from $n - 1$ to k.

For $k + l \leqslant n$, the transformation degenerates; the nontruncated constituent matrices constructed on h_2 drop out:

$$P_{2^n}^{(k)} F_{2^n} P_{2^n}^{(l)} = \left[\left(\prod_{i=n-1}^{n-l} (I_{2^{n-i-1}} \times S_2 \times P_{2^i}^{(k)}) \cdot \right. \right.$$

$$\left. \cdot [I_{2^{n-i-1}} \times ((I_2 \times P_{2^i}^{(k)}) D_{2^{i+1}}^{(i,\,k)})] \right) \cdot (I_{2^l} \times P_{2^{n-l-k}} \times S_{2^k}^T) \Big] M_{2^n}^{\text{in}}. \tag{2.94}$$

Figures 2.6 and 2.7 show transformation diagrams which illustrate Eqs. (2.93) and (2.94) for the cases $n = 5, k = 3, l = 3$ and $n = 5, k = 2, l = 2$, respectively. The numbers in the circles in this diagram show the power of the complex ex-

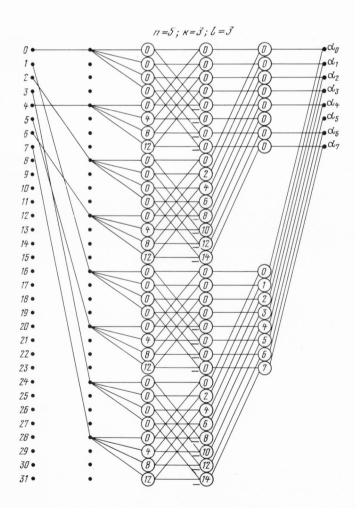

Fig. 2.6

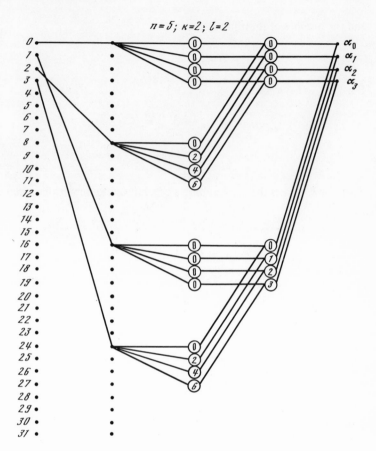

Fig. 2.7

ponential function exp $[i2\pi/32]$ which multiplies the element corresponding to the given position in the diagram.

What is the advantage of the truncation of the transformation in terms of the reduction in the number of operations? It can be seen from (2.93) and (2.94) and also from Figs. 2.6 and 2.7 that operations of three types must be carried out in the course of the transformation: addition (subtraction), multiplication, and transfer. We denote by T_{ad}, the time required to add (or subtract) two complex numbers, by T_{mu} the multiplication time, and by T_t the transfer time. We will ignore the time expended on the binary inversion since it can be carried out at the same time as the transfer. For $k + l \geqslant n$, the number of additions N_{ad} in step i of the transformation $(n - 1 \geqslant i \geqslant k)$ is

$$N_{\text{ad}}^{(i)} = 2^{n-i+k-1}, \tag{2.95}$$

and the number of multiplications is

$$N_{\text{mu}}^{(i)} = 2^{n-i+k-1} - 2^{n-i-1}. \tag{2.96}$$

For $k - 1 \geqslant i \geqslant n - l$,

$$N_{\text{ad}}^{(i)} = 2^n, \tag{2.97}$$

$$N_{\text{mu}}^{(i)} = 2^{n-1} - 2^{n-i-1}. \tag{2.98}$$

Then for $k + l \geqslant n$ the total numbers of additions and multiplications are

$$N_{\text{ad}} = \sum_{k}^{n-1} 2^{n-i+k-1} + 2^n (k + l - n) =$$
$$= 2^n (k + l + 1 - n) - 2^k, \tag{2.99}$$

$$N_{\text{mu}} = \sum_{k}^{n-1} 2^{n-i+k-1} - \sum_{n-l}^{k-1} 2^{n-i-1} - \sum_{k}^{n-1} 2^{n-i-1} + \sum_{n-l}^{k-1} 2^{n-1} =$$
$$= 2^n + 2^{n-1} (l + k - n) - (2^k + 2^l) + 1. \tag{2.100}$$

The number of transfer operations is 2^n. The total time required for the transformation can thus be estimated as

$$T^{(k,\,l)} = T_{\text{ad}} (2^n (k + l - n + 1) - 2^k) +$$
$$+ T_{\text{mu}}(2^n + 2^{n-1} (l + k - n) - (2^k + 2^l) + 1) +$$
$$+ T_{\text{t}} \cdot 2^n, \quad k + l \geqslant n. \tag{2.101}$$

The time required for a nontruncated transformation ($k = l = n$), on the other hand, is†

$$T^{(n,\,n)} = T_{\text{ad}} n \cdot 2^n + T_{\text{mu}} [(n/2 - 1) 2^n + 1] + T_{\text{t}} \cdot 2^n. \tag{2.102}$$

For $k + l \leqslant n$ the transformation time is

$$T^{(k,\,l)} = T_{\text{ad}} (2^{l+k} - 2^k) + T_{\text{mu}}(2^{l+k} - 2^l + 2^k + 1) +$$
$$+ T_{\text{t}} \cdot 2^{l+k}. \tag{2.103}$$

It is useful to estimate separately the advantage in terms of the number of additions and multiplications because of the truncation of the transformation. For $k + l \geqslant n$

†This is a refinement of the rough estimate above.

$$N_{ad}^{(k,\,l)}/N_{ad}^{(n,\,n)} = (k + l - n + 1 - 2^{k-n})/n, \qquad (2.104)$$

$$N_{mu}^{(k,\,l)}/N_{mu}^{(n,\,n)} = (k + l - n + 2 - 2^{k+1-n} - 2^{l+1-n} + 2^{-n})/(n - 2 + 2^{1-n}). \qquad (2.105)$$

For $k + l \leqslant n$,

$$N_{ad}^{(k,\,l)}/N_{ad}^{(n,\,n)} = (2^{l+k-n} - 2^{k-n})/n, \qquad (2.106)$$

$$N_{mu}^{(k,\,l)}/N_{mu}^{(n,\,n)} = (2^{l+k+1-n} - 2^{l+1-n} - 2^{k+1-n} + 2^{1-n})/(n - 2 + 2^{1-n}). \qquad (2.107)$$

The values of $N_{ad}^{(k,l)}/N_{ad}^{(n,n)}$ and $N_{mu}^{(k,l)}/N_{mu}^{(n,n)}$ are shown in Tables 2.2 and 2.3, respectively, for $n = 10$ and several values of k and l.

Joint DFT Algorithms. It follows from the properties of the DFT that the DFT of sequences of real numbers is redundant by a factor of 2: The Fourier coefficients with indices which add to the length of the sequence are complex conjugates (see Appendix I):

$$\alpha_s = \alpha_{N-s}^*. \qquad (2.108)$$

Then it is sufficient to calculate α_s for only $s = 0, 1, \ldots, N/2$; the other α_s can be found from (2.108) without going through the calculations. This circumstance can be exploited to reduce by a factor of about 2 the number of operations required to calculate the DFT of sequences of real numbers.

This advantage can be realized in two ways: by a joint transformation of two sequences; and by breaking up one sequence, having an even number of terms, into two subsequences, followed by a joint transformation of these subsequences and a conversion of the result to correspond to the entire sequence.

TABLE 2.2

	$N_{ad}^{(k,\,l)}/N_{ad}^{(n,\,n)}$					
l	$k = 5$	6	7	8	9	10
5	0.1	0.19	0.29	0.375	0.45	0.5
6	0.2	0.29	0.39	0.475	0.55	0.6
7	0.3	0.39	0.49	0.575	0.65	0.7
8	0.4	0.49	0.59	0.675	0.75	0.8
9	0.5	0.58	0.69	0.775	0.85	0.9
10	0.6	0.69	0.79	0.875	0.95	1

TABLE 2.3

l	$N_{\text{mu}}^{(k,\,l)}/N_{\text{mu}}^{(n,\,n)}$					
	$k = 5$	6	7	8	9	10
5	0.23	0.35	0.46	0.55	0.61	0.61
6	0.35	0.47	0.58	0.67	0.73	0.73
7	0.46	0.58	0.69	0.78	0.84	0.84
8	0.55	0.67	0.78	0.87	0.93	0.93
9	0.61	0.73	0.84	0.93	1	1
10	0.61	0.73	0.84	0.93	1	1

We will now consider the first method since the second ultimately reduces to the first.

Let $\{a_k\}$ and $\{b_k\}$ be two sequences of real numbers of length $N(k = 0, 1, 2, \ldots, N - 1)$. We form the sequence

$$c_k = a_k + ib_k \tag{2.109}$$

and find its DFT:

$$
\begin{aligned}
\gamma_s &= \frac{1}{\sqrt{N}} \sum_{k=0}^{N-1} c_k \exp\left(i2\pi \frac{ks}{N}\right) = \\
&= \frac{1}{\sqrt{N}} \sum_{k=0}^{N-1} (a_k + ib_k) \exp\left(i2\pi \frac{ks}{N}\right) = \alpha_s + i\beta_s = \\
&= \alpha_s^{\text{re}} - \beta_s^{\text{im}} + i\,(\alpha_s^{\text{im}} + \beta_s^{\text{re}}),
\end{aligned}
\tag{2.110}
$$

where

$$
\begin{aligned}
\alpha_s &= \frac{1}{\sqrt{N}} \sum_{k=0}^{N-1} a_k \exp\left(i2\pi \frac{ks}{N}\right), \\
\beta_s &= \frac{1}{\sqrt{N}} \sum_{k=0}^{N-1} b_k \exp\left(i2\pi \frac{ks}{N}\right),
\end{aligned}
\tag{2.111}
$$

and the superscripts "re" and "im" denote the real and imaginary parts of the corresponding numbers. By virtue of (2.108) we obviously have

$$\gamma_{N-s} = \alpha_{N-s} + i\beta_{N-s} = \alpha_s^* + i\beta_s^*. \tag{2.112}$$

Thus

$$\gamma_s + \gamma_{N-s} = (\alpha_s + \alpha_s^*) + i\,(\beta_s + \beta_s^*) = 2\alpha_s^{re} + i2\beta_s^{re},$$
$$\gamma_s - \gamma_{N-s} = (\alpha_s - \alpha_s^*) + i\,(\beta_s - \beta_s^*) = -\,2\beta_s^{im} + i2\alpha_s^{im}, \quad (2.113)$$

i.e.,

$$\alpha_s = (\gamma_s + \gamma_{N-s}^*)/2 = (\gamma_s^{re} + \gamma_{N-s}^{re})/2 + i\,(\gamma_s^{im} - \gamma_{N-s}^{im})/2,$$
$$\beta_s = -\,i\,(\gamma_s - \gamma_{N-s}^*)/2 = (\gamma_s^{im} + \gamma_{N-s}^{im})/2 - i\,(\gamma_s^{re} - \gamma_{N-s}^{re})/2. \quad (2.114)$$

Equations (2.114) show how to find the Fourier coefficients of the sequences $\{a_k\}$ and $\{b_k\}$ from the result of a transformation of the joint sequence $\{a_k + ib_k\}$ after carrying out $N + 2$ additions of real numbers [the calculations from (2.114) need be carried out for only $s = 0, 1, \ldots, N/2$; the other α_s and β_s can be found from (2.108)].

This calculation procedure is illustrated by Fig. 2.8. This joint-DFT method is useful, for example, in the transformation of two-dimensional arrays, in which case adjacent pairs of rows in the array are convenient as $\{a_k\}$ and $\{b_k\}$.

In a transformation of one-dimensional real arrays, the second method for reducing the number of operations is more convenient, as can be seen from the following discussion. Let $\{a_k\}$ be a sequence of real numbers of length $2N$ ($k = 0, 1, 2, \ldots, 2N - 1$). We are to find its DFT,

$$\alpha_s = \frac{1}{\sqrt{2N}} \sum_{k=0}^{2N-1} a_k \exp\left(i2\pi\,\frac{ks}{N}\right). \quad (2.115)$$

Distinguishing the even and odd terms in the sum in (2.115), we can write

$$\alpha_s = \frac{1}{\sqrt{2N}} \left\{ \sum_{k=0}^{N-1} a_{2k} \exp\left(i2\pi\,\frac{2ks}{2N}\right) + \sum_{k=0}^{N-1} a_{2k+1} \exp\left[i2\pi\,\frac{(2k+1)\,s}{2N}\right] \right\} =$$
$$= \frac{1}{\sqrt{2N}} \left\{ \sum_{k=0}^{N-1} a_{2k} \exp\left(i2\pi\,\frac{ks}{N}\right) + \right.$$
$$\left. + \left[\sum_{k=0}^{N-1} a_{2k+1} \exp\left(i2\pi\,\frac{ks}{N}\right) \right] \exp\left(i\pi\,\frac{s}{N}\right) \right\}. \quad (2.116)$$

The DFT of the entire sequence $\{a_k\}$ can be found by calculating the DFTs of the two subsequences of this sequence, which consists of the even and odd terms, respectively, and of the original sequence and by then summing the results in accordance with

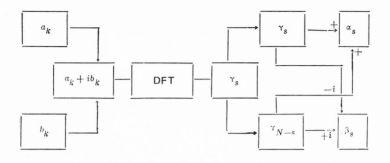

Fig. 2.8

$$\alpha_s = \frac{1}{\sqrt{2}} [\alpha_s^e + \exp{(i\pi s/N)}\,\alpha_s^o], \qquad (2.117)$$

where α_s^e and α_s^o are the DFTs of $\{a_{2k}\}$ and $\{a_{2k+1}\}$, respectively. The latter DFTs can be found by using the first method, the joint-DFT method, and (2.114):

$$\alpha_s = \frac{1}{2\sqrt{2}} \left\{ (\gamma_s + \gamma_{N-s}^*) - i\exp\left(i\frac{\pi s}{N}\right)[(\gamma_s - \gamma_{N-s}^*)] \right\} =$$
$$= \frac{1}{2\sqrt{2}} \left\{ (\gamma_s^{re} + \gamma_{N-s}^{re}) + i(\gamma_s^{im} - \gamma_{N-s}^{im}) + \right.$$
$$\left. + \exp\left(i\pi\frac{s}{N}\right)[(\gamma_s^{im} + \gamma_{N-s}^{im}) - i(\gamma_s^{re} - \gamma_{N-s}^{re})] \right\}, \qquad (2.118)$$

where

$$\gamma_s = \frac{1}{\sqrt{N}} \sum_{k=0}^{N-1} (a_{2k} + ia_{2k+1}) \exp\left(i\,2\pi\,\frac{ks}{N}\right) \qquad (2.119)$$

and $s = 0, 1, 2, \ldots, N$. The values of α_s for other values of s are found from (2.108).

A comparison of (2.114) and (2.118) shows that the second method requires more complicated additional calculations (aside from the additions, there are also complex multiplications).

These algorithms for joint Fourier transformations are easily inverted, and it is simple to find algorithms for calculating the DFTs of sequences in which pairs of elements which are symmetric with respect to the center of the sequence are complex-conjugate numbers, as in (2.108). Returning to Fig. 2.8, we easily see that if we have two sequences $\{\alpha_s\}$ and $\{\beta_s\}$ such that $\alpha_s = \alpha_{N-s}^*$, $\beta_s = \beta_{N-s}^*$, we can use them to form sequences γ_s in accordance with the rule

$$\gamma_s = \alpha_s + i\beta_s, \quad \gamma_{N-s}^* = \alpha_s - i\beta_s,$$
$$s = 0, 1, 2, \ldots, N/2. \qquad (2.120)$$

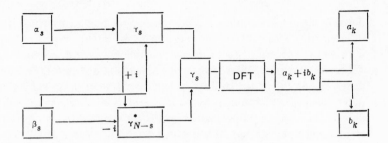

Fig. 2.9

Then the real part of the Fourier transformation of this sequence is the Fourier transformation $\{\alpha_s\}$, while the imaginary part is the Fourier transformation $\{\beta_s\}$. The diagram corresponding to this algorithm is shown in Fig. 2.9.

In precisely the same way, if we have a single sequence of length $2N\{\alpha_s = \alpha_{2N-s}^*\}$, we can use it to form two sequences,

$$\alpha_s^e = 1/2\,(\alpha_s + \alpha_{N-s}^*), \quad \alpha_{N-s}^e = \alpha_s^{e*},$$

$$\alpha_s^o = 1/2\,\exp\left(-i\pi\,\frac{s}{N}\right)[\alpha_s - \alpha_{N-s}^*], \quad \alpha_{N-s}^o = \alpha_{N-s}^{o*}, \quad (2.121)$$

$$s = 0,\, 1,\, 2,\, \ldots,\, N/2,$$

and then we proceed as in the preceding paragraph for two separate sequences.

The joint-transformation algorithms can also be used to calculate the two-dimensional transformations of real or complex-conjugate arrays. These transformations are carried out as two one-dimensional transformations. In the transformation of two-dimensional arrays of real numbers, the first Fourier transformation can be carried out by joining the DFTs of pairs of rows of the array, and the second Fourier transformation of the resulting array of complex numbers is carried out only up to the halfway point in terms of the number of columns. The second half is found as the complex conjugate of the first half, without the need for calculations. In the transformation of complex-conjugate arrays it is necessary to work in the opposite order: The first Fourier transformation is carried out only halfway through the array; the results are then supplemented with the numbers which are the complex conjugates of the results of the first transformation. Then the second Fourier transformation is carried out by the algorithm described above for the joint transformation of two sequences with elements which are pairs of complex conjugates.

Section 3. Analysis of Analog Devices for Hologram Recording and Reconstruction

Description of Recording Apparatus and Devices for Hologram Recording and Reconstruction. In order to produce a synthetic hologram, we must transform the array of numbers which are found through calculations and which describe the hologram into a physical object (the hologram) which is capable of interacting with light to restore the image for the observer. For this procedure we need special devices for recording holograms at the computer output. These devices can differ in design and can perform different transformations, depending on the particular discrete representation of the hologram in the computer. Since we are dealing with the representation of holograms as three-dimensional readings, we will analyze only those devices which record these readings on a physical carrier and which determine their interpolation.

Various physical carriers can be used to record holograms: photographic materials, thermoplastics, and liquid crystals, among others. From the standpoint of physical limitations, there are no fundamental differences among these materials. For definitiveness we will speak in terms of photographic materials exclusively since these materials were used in all the experiments described below.

Since there are technical difficulties in extracting data on all the readings of the hologram simultaneously from the computer and since the recording devices have their technological limitations, these devices are constructed as sequential-action units, which record the readings one by one in a scanning process. They usually consist of a device for scanning the photosensitive material and an illuminator controlled by a modulator (Fig. 3.1). The modulator takes the data from the computer to determine the amount of light with which a given part of the photosensitive material (photographic film, plates, etc.) is exposed. The system consisting of the modulator, the illuminator, and the photosensitive material thus transforms the digital signal from the computer into an analog signal, represented by the degree of blackening of the photographic material. The illuminator usually acts on spots of finite size on the photographic material, so it performs an analog interpolation of the readings of the recorded hologram.

Systems of this type have certain limitations:

1. It is not possible to record both parts of a complex number† which correspond to the amplitude and phase of the light, on the same region on the carrier. The recorded value must be nonnegative.
2. The recording media are nonlinear; that is, the physical parameter which is used for the recording (a transmission coefficient, a reflection coefficient, etc.) varies in a nonlinear way with the quantity being recorded.

---

†Systems have recently been introduced which can reproduce color images in multilayer photographic materials from a light signal. Systems of this type can be used to record complex numbers, by exposing one sensitive layer of the material with a signal proportional to the amplitude and exposing another layer with a signal proportional to the phase and then developing this second layer in a manner designed to produce a phase relief. There have been reports of the use of such systems to record computer-generated holograms [17–19].

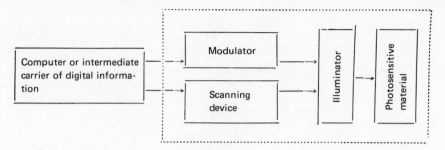

Fig. 3.1

3. The recording systems have a finite resolution. Only a finite number of linearly independent quantities can be recorded on a unit area of the carrier.

These limitations, which are imposed by the very nature of the recording systems, can be described mathematically in a first approximation by representing the recording systems as a nonlinear unit with an infinite resolution which is connected in series with a linear unit which describes the finite resolution of the system and by assuming that the signals can take on only real values.

Methods for Recording Complex Quantities. The problem of recording complex quantities is the first problem with which we must deal if we wish to synthesize holograms. Since the physical carrier can convey only a real quantity per element of resolution, other elements must be used in order to record complex signals. The minimum number of resolution elements required is two per complex value.

Lohmann, Paris, Brown *et al.* [20-26] have proposed a method for recording complex quantities in devices which are capable of conveying only two states. In this method of "binary holography," the modulus of the complex quantity is conveyed by the dimensions of the aperture in the carrier which transmits the light, while the phase is conveyed by the position of the aperture with respect to a fixed grid. The grid spacing is set at a value such that the phase difference between the rays from adjacent edges of the grid at the observation angle with respect to the plane of the hologram is 2π. For this method the number of degrees of freedom (resolution elements) must exceed the number of readings of the hologram by a factor of about NM, where N is the number of discrete amplitude levels, and M is the number of discrete levels of the phase of the recorded hologram.

Lee [27-29] has suggested separate transmission of the real and imaginary parts of the complex quantity and the separation of positive and negative values. The purpose here is to make better use of the dynamic range of the recording medium. The procedure is as follows: Let F_r and F_i be the real and imaginary parts of a reading of the hologram. We determine the nonnegative real quantities

$$F_1 = \begin{cases} F_r & F_r \geqslant 0, \\ 0 & F_r < 0, \end{cases} \qquad F_3 = \begin{cases} -F_r & F_r < 0, \\ 0 & F_r \geqslant 0, \end{cases}$$

$$\text{(3.1)}$$

$$F_2 = \begin{cases} F_i & F_i \geqslant 0, \\ 0 & F_i < 0, \end{cases} \qquad F_4 = \begin{cases} -F_i & F_i < 0, \\ 0 & F_i \geqslant 0. \end{cases}$$

Then each reading of the hologram can be written in the form

$$F(r, s) = F_1(r, s) - F_3(r, s) + iF_2(r, s) - iF_4(r, s). \quad \text{(3.2)}$$

Four resolution elements on the carrier are used to convey each set of four non-negative values (F_1, F_2, F_3, F_4), two of which are always zero (Fig. 3.2). These four resolution elements are arranged in succession in one direction. When the hologram is illuminated by coherent light, the phase difference between the rays which are transmitted through two adjacent resolution elements at an angle $\theta = \arccos(\lambda/4\Delta\xi)$ (λ is the wavelength and $\Delta\xi$ is the spacing of the elements in the hologram) is $\pi/2$, so that condition (3.2) is satisfied in the reconstruction. The number of degrees of freedom required of the physical hologram is thus four times the number of readings of the mathematical hologram.

A way to reduce this ratio to $1 : 3$ has been proposed [30]. In this approach the complex number is represented by nonnegative components in a two-dimensional simplex (Fig. 3.3b), in contrast with Lee's method in which the complex number is represented as a vector with nonnegative components in "biorthogonal" coordinates (Fig. 3.3a). In this approach, the complex number is conveyed by three successive resolution elements. The angle at which the reconstruction is carried out is such that the phase between the rays transmitted through any two adjacent elements of the hologram is $120°$.

If the restriction that only nonnegative numbers be transmitted is dropped, it is possible to propose another modification of Lee's method: to use two adjacent elements of the physical hologram to convey real and imaginary parts

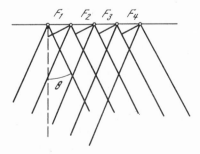

Fig. 3.2

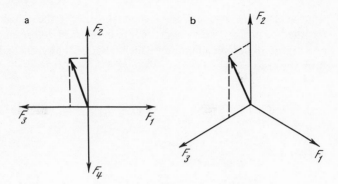

Fig. 3.3

of the holograms. Then the image is reconstructed at an angle for which the phase difference between the rays transmitted through adjacent elements is $\pi/2$. Since the path difference between even elements (or between odd elements) is π, it is necessary to change the signs of the real and imaginary parts of all the odd (or all the even) readings of the mathematical hologram before the recording.

In yet another modification of Lee's method, the readings conveying the real and imaginary values of the readings of a hologram with nonnegative values are in a more symmetric arrangement. In contrast with the preceding case, it is necessary to use two pairs of adjacent elements, in adjacent rows, to convey the real and imaginary parts of the hologram (instead of two adjacent elements on the carrier). Each element of each pair is involved in conveying positive and negative values of the real and imaginary parts of the hologram reading, respectively. The image is formed in a direction which makes angles with the directions along and across the rows of the physical hologram such that the path difference between the adjacent pairs of elements is $\pi/2$, while that between the elements of the same pair is π.

A method for putting the object in a symmetric form in order to obtain a purely real mathematical hologram has been proposed [31]. In practice, this result can be achieved by using either the real or the imaginary parts of the complex mathematical hologram. To prevent the superposition of the object on itself in the reconstruction from such a hologram, the object must be supplemented with an empty field of size equal to or larger than that of the object. If the real part is taken after a one-dimensional Fourier transformation, this procedure corresponds to a quadrupling of the object; if the real part is taken after a two-dimensional Fourier transformation, there is a doubling of the object. The ratios of the number of readings of the mathematical hologram and of the recorded hologram are 1:4 and 1:2, respectively.

We can show that all these methods for recording complex quantities are equivalent to the introduction of a spatial carrier in the hologram before it is recorded by analogy with the procedure in physical holography. Let $\{F(r,s)\}$ be a matrix of complex numbers which describe the synthetic hologram. We form the array

$$\Phi(k,l) = \text{re}\left\{F(r,s)\exp\left(i\,2\pi\frac{rk_1+k_2}{a}\right)\exp\left(i2\pi\frac{sl_1+l_2}{b}\right)\right\},\quad (3.3)$$

where

$$k = rk_1 + k_2;\quad l = sl_1 + l_2;\quad k_2 = 0,1,\ldots,k_1-1;$$
$$l_2 = 0,1,\ldots,l_1-1;\quad r = 0,1,\ldots,N_x-1;\quad s = 0,1,\ldots$$
$$\ldots,N_y-1;$$

and k_1, l_1, a, and b are constants.

The complex numbers $\Phi(k,l)$ can be thought of as the readings of a hologram in which spatial carriers with frequencies $1/a$ and $1/b$ have been inserted.

The double indexing (r and k_2; s and l_2) explicitly reflects the introduction of intermediate readings in the formation of the spatial carrier. The numbers of intermediate readings are $k_1 - 1$ and $l_1 - 1$. Equation (3.3) means that the values of the hologram are interpolated in a stepwise manner at these intermediate readings:

$$F(k,l) = F(r,s). \tag{3.4}$$

Let us consider some particular cases.

1. $k_1 = 2, l_1 = 1, a = 4, b = 1$. In this case the period of the spatial carrier along the coordinate r is two readings of the hologram $F(r,s)$, and this period is equal to four digitization intervals, i.e., four times the reading along the k scale. Here

$$\Phi(k,l) = \text{re}\left\{F(r,s)\exp\left(i2\pi\frac{2r+k_2}{4}\right)\right\} =$$
$$= \text{re}\left\{F(r,s)\exp i\left(\pi r + \frac{\pi}{2}k_2\right)\right\} = \text{re}\{(-1)^r i^{k_2}F(r,s)\}. \tag{3.5}$$

Thus

$$\Phi(k,l) = \begin{cases} \text{re}\,\{F(r,s)\} & k_2 = 0 \quad \text{even } r, \\ -\text{re}\,\{F(r,s)\} & k_2 = 0 \quad \text{odd } r, \\ \text{im}\,\{F(r,s)\} & k_2 = 1 \quad \text{even } r, \\ -\text{im}\,\{F(r,s)\} & k_2 = 1 \quad \text{odd } r, \end{cases} \tag{3.6}$$

This is the case (described above) of the recording of the real and imaginary parts at pairs of successive resolution elements of the physical hologram [see Fig. 3.4, which shows the grid on the physical hologram and the positions of the real (re) and imaginary (im) parts of the readings].

2. $k_1 = 2, l_1 = 2, a = 4, b = 2$. In this case the period of the spatial carrier along the coordinate l is equal to one reading of the hologram $F(r, s)$ and to two digitization intervals along this coordinate:

$$\Phi(k, l) = \text{re} \left\{ F(r, s) \exp \left[i \, 2\pi \left(\frac{2r + k_2}{4} + \frac{2s + l_2}{2} \right) \right] \right\} =$$
$$= \text{re} \{ (-1)^{r+l_2} (i)^{k_2} F(r, s) \}. \tag{3.7}$$

Figure 3.5 shows the arrangement of the recorded quantities among the resolution elements of the carrier for this case.

3. If the period of the spatial carrier is set equal to one reading of the original hologram and four digitization intervals along k, we have the Lee method. Formally, this case can be described by introducing an indexing function with respect to k:

$$k = 4r + 2r_1 + r_2,$$

where

$$r = 0, 1, \ldots, N_x - 1; r_1, r_2 = 0, 1,$$

such that

$r = 0, k_2 = 0$	$r = 0, k_2 = 1$	$r = 1, k_2 = 0$	$r = 1, k_2 = 1$	$r = 2, k_2 = 0$
$k = 0$	$k = 1$	$k = 2$	$k = 3$	$k = 4$
re_0	im_0	$-\text{re}_1$	$-\text{im}_1$	re_2

Fig. 3.4

$k_2 = 0$ $l_2 = 0$ $r = 0$ $s = 0$	$k_2 = 1$ $l_2 = 0$ $r = 0$ $s = 0$	$k_2 = 0$ $l_2 = 0$ $r = 1$ $s = 0$	$k_2 = 1$ $l_2 = 0$ $r = 1$ $s = 0$
re $F(0,\ 0)$	$-$im $F(0,0)$	$-$re $F(1,\ 0)$	im $F(1,\ 0)$
$k_2 = 0$ $l_2 = 1$ $r = 0$ $s = 0$	$k_2 = 1$ $l_2 = 1$ $r = 0$ $s = 0$	$k_2 = 0$ $l_2 = 1$ $r = 1$ $s = 0$	$k_2 = 1$ $l_2 = 1$ $r = 1$ $s = 0$
$-$re $F(0,\ 0)$	im $F(0,\ 0)$	re $F(1,\ 0)$	$-$im $F(1,0)$
$k_2 = 0$ $l_2 = 0$ $r = 0$ $s = 1$	$k_2 = 1$ $l_2 = 0$ $r = 0$ $s = 1$	$k_2 = 0$ $l_2 = 0$ $r = 1$ $s = 1$	$k_2 = 1$ $l_2 = 0$ $r = 1$ $s = 1$
re $F(0,\ 1)$	$-$im $F(0,1)$	$-$re $F(1,\ 1)$	im $F(1,\ 1)$
$k_2 = 0$ $l_2 = 1$ $r = 0$ $s = 1$	$k_2 = 1$ $l_2 = 1$ $r = 0$ $s = 1$	$k_2 = 0$ $l_2 = 1$ $r = 1$ $s = 1$	$k_2 = 1$ $l_2 = 1$ $r = 1$ $s = 1$
$-$re $F(0,\ 1)$	im $F(0,\ 1)$	re $F(1,\ 1)$	$-$im $F(1,\ 1)$

Fig. 3.5

$$\Phi\left(k, l\right) = \mathrm{re}\left\{F\left(r, s\right)\exp\left(i\,2\pi\,\frac{4r + 2r_1 + r_2}{4}\right)\right\} =$$

$$= \mathrm{re}\left\{(-1)^{r_1}\,(i)^{r_2}\,F\left(r, s\right)\right\}, \tag{3.8}$$

i.e.,

$$\Phi\left(k, l\right) = \begin{cases} \mathrm{re}\left\{F\left(r, s\right)\right\}, & r_1 = 0, & r_2 = 0, \\ \mathrm{im}\left\{F\left(r, s\right)\right\}, & r_1 = 0, & r_2 = 1, \\ -\mathrm{re}\left\{F\left(r, s\right)\right\}, & r_1 = 1, & r_2 = 0, \\ -\mathrm{im}\left\{F\left(r, s\right)\right\}, & r_1 = 1, & r_2 = 1. \end{cases} \tag{3.9}$$

Nonlinearity of the Recording System. Let T be a parameter which relates the amplitude of the light incident on the recording medium to the amplitude of the light which has reacted with the medium:

$$A_{\text{out}} = TA_{\text{in}}. \tag{3.10}$$

In general, this parameter can be complex. For example, the modulus of T may describe the transparency of the photographic carrier, while the phase of T may describe the optical thickness of the carrier. In practice, however, it is difficult to control the modulus and phase of T simultaneously, so it is necessary either to "sandwich" together separate amplitude and phase recording media or to restrict the recording to the modulus of T alone. Here we will discuss only the latter case:

$$A_{\text{out}} = |T|\,A_{\text{in}}. \tag{3.11}$$

We denote by u the signal which is transformed into $|T|$ in the recording system:

$$|T| = T\left(u\right) \geqslant 0, \quad u \in [0, 1]. \tag{3.12}$$

The function $T(u)$ is generally nonlinear, so the signal must also be a nonlinear function of the recorded quantity to offset the nonlinearity of $T(u)$. Since, in accordance with the principles of hologram synthesis, the quantity to be recorded is proportional to the light amplitude at the exit from the hologram, the variation of the signal u with the recorded quantity $F_b(r, s)$ must be of the type

$$u = T^{-1}\left(F_b\left(r, s\right) + F_0\right), \tag{3.13}$$

where T^{-1} is the inverse of $T(u)$ and F_0 satisfies the condition

$$F_b\left(r, s\right) + F_0 \geqslant 0. \tag{3.14}$$

Let us consider one of the most common cases, in which the hologram is re-corded on a photographic carrier and the control signal is proportional to the blackening density:

$$u = \frac{\ln (T/T_{\max})}{\ln (T_{\min}/T_{\max})}, \tag{3.15}$$

where $T_{\min}$ and $T_{\max}$ are the minimum and maximum values of the transmission coefficient of the material, i.e.,

$$T/T_{\max} = (T_{\min}/T_{\max})^u = \exp (-u \ln k), \tag{3.16}$$

where $k = T_{\max}/T_{\min}$.

For T to be a linear function of $F_{\hat{b}}(r, s)$ in this case,

$$T/T_{\max} = k_1 F_{\hat{b}}(r, s) + k_2, \tag{3.17}$$

the condition

$$u = - \{\ln [k_1 F_{\hat{b}}(r, s) + k_2]\}/\ln k \tag{3.18}$$

must hold. The constants k_1 and k_2 in the linear relation in (3.17) are governed by the boundary values of $F_{\hat{b}}$ and T from the equations†

$$1/k = - k_1 | F_{\hat{b}\,\min}| + k_2, \quad 1 = k_1 F_{\hat{b}\,\max} + k_2, \tag{3.19}$$

Using

$$\varkappa = F_{\hat{b}\,\max} /| F_{\hat{b}\,\min}|, \tag{3.20}$$

we then find

$$k_1 = \frac{1}{| F_{\hat{b}\,\min}|} \frac{k-1}{k(\varkappa+1)}, \quad k_2 = \frac{k+\varkappa}{k(\varkappa+1)}. \tag{3.21}$$

Writing

$$f(r, s) = F_{\hat{b}}(r, s)/(F_{\hat{b}\,\max} + | F_{\hat{b}\,\min}|), \tag{3.22}$$

we finally find

---

†We recall that $F_{\hat{b}}$ can take on both positive and negative values, so that $|F_{\hat{b}\max}|$ is the largest negative value of $F_{\hat{b}}$.

$$u = - \left[\ln\left(\frac{k-1}{k} f + \frac{\varkappa + k}{k(\varkappa + 1)} \right) \right] \Big/ \ln k. \tag{3.23}$$

This equation describes the nonlinear preliminary "distortion" to which the matrix of numbers describing the mathematical hologram should be subjected before the hologram is recorded on the photographic carrier in a device where the blackening density is linearly related to the signal.

Finite Aperture of the Photographic System. We assume that the nonlinearity of the recording device has been corrected by the appropriate "distortion" of the signal before the recording. Therefore the process of converting the matrix of numbers into a physical hologram in the photographic system can be described by

$$\Gamma_p(v_x, v_y) = \sum_{r=0}^{N_x-1} \sum_{s=0}^{N_y-1} [pF_{\hat{\delta}}(r, s) + q] \times$$
$$\times H(v_x + v_{0x} - r\Delta v_x, v_y + v_{0y} - s\Delta v_y), \tag{3.24}$$

where p and q are constants which arise as a result of the correction of the recording nonlinearity, Δv_x and Δv_y are the digitization steps along the v_x and v_y axes in the recording device, v_{0x} and v_{0y} are constants which are governed by the geometry of the hologram in the reconstruction apparatus, and $H(v_x, v_y)$ is the point scattering function of the recording device. For example, this function describes the intensity distribution in the light spot if a device with a traveling beam is used for the recording.

Equation (3.24) can be rewritten

$$\Gamma_p(v_x, v_y) = \Pi(v_x, v_y) \Big[H(v_x, v_y) \circledast$$
$$\circledast \sum_{r=-\infty}^{\infty} \sum_{s=-\infty}^{\infty} [pF_{\hat{\delta}}(r, s) + q] \delta(v_x + v_{0x} - r\Delta v_x) \times$$
$$\times \delta(v_y + v_{0y} - s\Delta v_y) \Big], \tag{3.25}$$

where $\circledast$ means a convolution, and

$$\Pi(v_x, v_y) = \begin{cases} 1 & \text{for } -v_{xm} \leqslant v_x \leqslant v_{xm}, \\ & \quad\ -v_{ym} \leqslant v_y \leqslant v_{ym}, \\ 0 & \text{otherwise} \end{cases} \tag{3.26}$$

is a unit function specified on the rectangle $(-v_{xm}, v_{xm}, -v_{ym}, v_{ym})$.

With the equation written in this form, the mosaic multiplication of the hologram during its recording can be taken into account [if $v_{x1} > v_{x0}$,

$v_{x2} > (N_x - 1)\,\Delta v_x - v_{x0}; v_{y1} > v_{y0}, v_{y2} > (N_y - 1)\,\Delta v_y - v_{y0}]$ and the effect of the function $H(v_x, v_y)$ on the reconstruction of the object can be estimated by making use of the convolution theorem from Fourier transform theory. According to this theorem, the result of the inverse Fourier transformation of the hologram $\Gamma_p(v_x, v_y)$, carried out in the reconstruction, can be written

$$b_b(x, y) = \iint\limits_{-\infty}^{\infty} \Gamma_p(v_x, v_y) \exp\{i2\pi(v_x x + v_y y)\}\, dx\, dy =$$

$$= \pi(x, y) \circledast \left[h(x, y) \iint\limits_{-\infty}^{\infty} \left(\sum_r \sum_s [pF_{\hat{b}}(r, s) + q] \times \right. \right.$$

$$\times\, \delta(v_x + v_{0x} - r\Delta v_x)\, \delta(v_y + v_{0y} - s\Delta v_y)\bigg) \times$$

$$\times \exp[-i2\pi(v_x x + v_y y)]\, dx\, dy \bigg] = \pi(x, y) \circledast \bigg\{ h(x, y) \times$$

$$\times \left[\frac{p}{\Delta v_x \Delta v_y} \sum_k \sum_l \sum_m \sum_n \hat{b}\left(m\Delta x + \frac{k}{\Delta v_x}, n\Delta y + \frac{l}{\Delta v_y} \right) \times \right.$$

$$\times\, \delta\left(x - m\Delta x - \frac{k}{\Delta v_x}, y - n\Delta y - \frac{l}{\Delta v_y} \right) +$$

$$+ \frac{q}{\Delta v_x \Delta v_y} \sum_k \sum_l \delta\left(x + \frac{k}{\Delta v_x}, y + \frac{l}{\Delta v_y} \right) \bigg] \bigg\}, \tag{3.27}$$

where

$$\pi(x, y) = \iint\limits_{-\infty}^{\infty} \Pi(v_x, v_y) \exp\{-i2\pi(v_x x + v_y y)\}\, dv_x dv_y =$$

$$= 4v_{xm}v_{ym}\,(\text{sinc}\ 2\pi v_{xm}x)\,(\text{sinc}\ 2\pi v_{ym}y), \tag{3.28}$$

$$h(x, y) = \iint\limits_{-\infty}^{\infty} H(v_x, v_y) \exp\{-i2\pi(v_x x + v_y y)\}\, dv_x dv_y. \tag{3.29}$$

In the simplest case of a purely rectangular aperture for the recording device,

$$H(v_x, v_y) = \begin{cases} 1, & |v_x| \leqslant \Delta v_x/2, \\ & |v_y| \leqslant \Delta v_y/2 \\ 0 & \text{otherwise} \end{cases} \tag{3.30}$$

we have

$$h(x, y) = \Delta v_x \Delta v_y\ \text{sinc}\ (2\pi\Delta v_x x)\ \text{sinc}\ (2\pi\Delta v_y y). \tag{3.31}$$

Since $\Delta v_x \ll v_{xm}$ and $\Delta v_y \ll v_{ym}$ (in the transmission of the entire hologram, their ratios are N_x and N_y, respectively), we see that $h(x,y)$ varies much more slowly than does $\pi(x,y)$. We can thus write, approximately,

$$b_b(x,y) \approx h_1(x,y) \left[p \sum_k \sum_l \sum_m \sum_n \hat{b} \left(m\Delta x + \frac{x}{\Delta v_x}, n\Delta y + \frac{l}{\Delta v_y} \right) \times \right.$$

$$\times \pi \left(x - m\Delta x + \frac{k}{\Delta v_x}, y - n\Delta y + \frac{l}{\Delta v_y} \right) +$$

$$\left. + q \sum_k \sum_l \pi \left(x + \frac{k}{\Delta v_x}, y + \frac{l}{\Delta v_y} \right) \right], \qquad (3.32)$$

where $h_1(x,y) = h(x,y)/\Delta v_x \Delta v_y$.

The image contains several diffraction orders (with indices k and l), which are masked by the function $h(x,y)$, which is the Fourier transform of the aperture of the writing element of the recording device. The masking is a consequence of the finite size of this aperture. The second term in brackets in (3.32) describes the "central beam," which results from the presence of a constant component in the recorded hologram.

Section 4. Experiments with Synthesized Holograms

Correction of Edge Shadow. As pointed out in Section 3, the finite aperture dimensions of the device used to record the hologram lead to an image that is shaded in accordance with the field of the masking function, which is proportional to the square of the Fourier transform of the pulsed response of the recording device. This shading can be corrected by multiplying the original amplitude distribution on the object by the function which is the inverse of the shadow function. In a practical visualization problem this shadow effect does not have to be corrected exactly. Even a crude approximation of the correcting function will be successful.

The shadow effect and its correction can be seen in Figs. 4.1–4.3. Figure 4.1 shows the image formed from a hologram synthesized from an object without a correction for shading. Because of the shadow effect here, the peripheral parts of the object have been lost from the image. Figure 4.2 shows the image formed from a hologram synthesized with a preliminary "distortion" of the original object. Here we can see an overcorrection. Figure 4.3 shows the function used for the correction; it is a parabolic function, which approximates the function which is the inverse of the frequency-contrast characteristic of the photographic recording system used. If the magnitude of this function is chosen appropriately, the shadow effect can be canceled satisfactorily (Fig. 4.4). Figure 4.4 was obtained by increasing the light amplitude at the edges of the original image by a factor of 8, while Fig.

Fig. 4.1

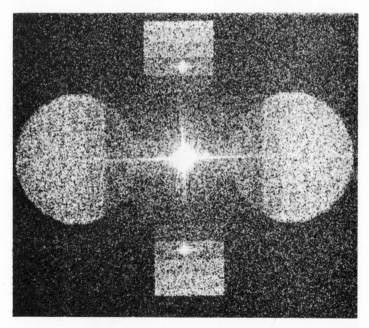

Fig. 4.2

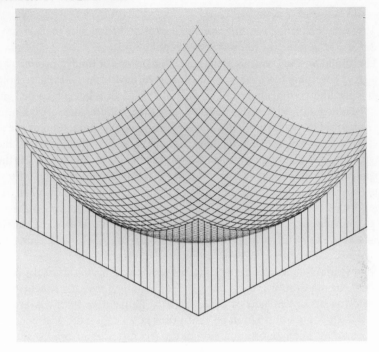

Fig. 4.3

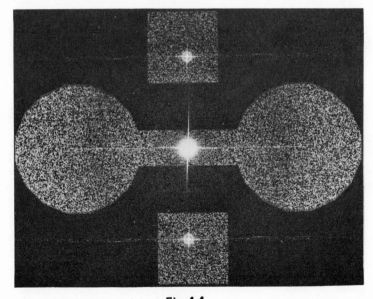

Fig. 4.4

4.2 was obtained through an increase by a factor of 30 and with an aperture of 12.5 × 12.5 μm for the photographic recording system.

Limitation on the Dynamic Range and Dimensions of the Hologram. One of the factors governing the quality of the images formed from synthesized holograms is the limited dynamic range of the hologram as it is being recorded and when the holograms are stored in a computer, if the computer memory is limited. Experiments with synthesized holograms and a computer simulation of the effects of a limitation (Section 8) show that the limited dynamic range of the holograms of specular objects† leads to a substantial image distortion. On the holograms of diffuse objects this limitation takes the form of speckle noise. Figure 4.5 illustrates the image distortion for specular objects; shown here is an image formed from a hologram in an optical system. The limitation on the hologram readings reduces the image to bare outlines. This result has a simple explanation: The dynamic range of the Fourier holograms of specular objects is very large because of the very large difference between the intensities of the low and high spatial frequencies in their Fourier spectrum. As a result of the limitation (and the digitization), the ratio of the low and high spatial frequencies is distorted toward an emphasis of the high spatial frequencies, with the result that, basically, only information on contours is being conveyed. This conclusion is confirmed by a computer simulation of the reconstruction from the synthesized hologram (Fig. 4.6).

A correct choice of the function to correct the recording nonlinearity can partially reduce the image distortion. In Fig. 4.7, for example, the image is still predominantly an outline, but low-frequency information is also being conveyed.

Holograms of diffuse objects are less affected by the limited dynamic range and the digitization. Figure 4.8 shows the image formed from the hologram of an object for which the phase of the reflection coefficient is specified as a pseudorandom quantity, taking on the values 0 and π with equal probabilities. This is done by multiplying, element by element, the original distribution of the specified field amplitude on the object by the equiprobable values ±1. Holograms of diffuse objects are far more homogeneous than those of specular objects. The information on the object which they carry is distributed over the entire area (Fig. 4.9), as in optical holograms recorded in an arrangement with diffuse illumination of the object. As a result, the dynamic range of the hologram is narrowed, and the effects of the limitation and quantization are restricted to the appearance of a speckle noise. There would be no such noise if the hologram could be recorded and reconstructed without distortion since the random

---

†We recall that a "specular" object is one for which the phase of the reflection (or transmission) coefficient is constant. In other words, the object is perfectly flat.

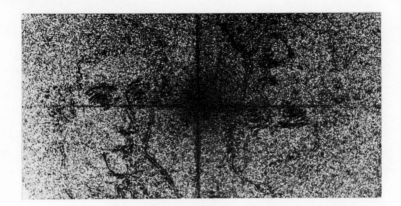

Fig. 4.5

Fig. 4.6

Fig. 4.7

Fig. 4.8

Fig. 4.9

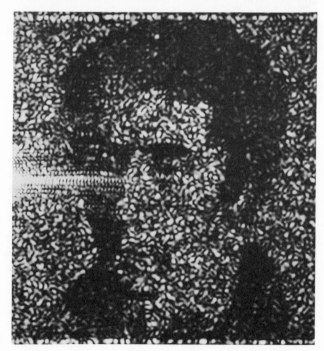

Fig. 4.10

phase specified on the object is not recorded in any way during the reconstruction; only the brightness is reconstructed.†

The limited dimensions of the holograms of diffuse objects have the same effect as a limitation of the dynamic range: The image is formed completely, but the speckle noise is intensified. This situation is illustrated by Fig. 4.10, which is an image formed from one-fourth the area of the hologram in Fig. 4.9. Comparison of Figs. 4.8 and 4.10 shows that the reduction of the hologram area used in the reconstruction results in an increase in the geometric dimensions of the speckle spots and thus a general increase in the image noise.

The speckle noise decreases as the hologram area increases. For a given resolution requirement at the object, an increase in the area of the hologram means an increase in its redundancy. One way to introduce a redundancy is to synthesize several holograms for various phase matrices and to form the image from all the holograms simultaneously, combined in a mosaic. This procedure is equivalent to expanding the spatial spectrum of the diffuser. As a result, the diffuser in a sense acquires a finer structure, and it is filtered out better when the image is observed. The same is true of a simple increase in the number of readings of the original object (cf. Figs. 4.11 and 4.8; the corresponding holograms have areas of 512×512 and 1024×1024 elements).

Introduction of a Spatial Carrier during the Recording of the Hologram. As mentioned in Section 3, all methods for recording holograms introduce a spatial carrier in one way or another. This carrier can be introduced before the holograms are synthesized, by displacing the object in the object plane, or immediately after the synthesis of the mathematical hologram, by multiplying the hologram by a complex exponential function with the appropriate spatial period.

The introduction of a spatial carrier of the maximum possible frequency through a displacement of the object is essentially the same as making the object symmetric. There are two ways to do this: by mirror quadrupling and by mirror doubling. Holograms of such objects are also symmetric (Figs. 4.12 and 4.13). In the quadrupling case the introduction of a spatial carrier requires a fourfold redundancy of the physical hologram with respect to the mathematical hologram and the object: There are four hologram readings corresponding to each reading of the object. The use of the symmetrizing procedure in hologram synthesis has programming advantages in certain cases in connection with the organization of the data storage in the computer [32, 33], but the introduction of a spatial carrier after the synthesis of a mathematical hologram is more natural and more convenient, especially since this step can in principle be carried out at the same

---

†We should point out that the noise visible in Fig. 4.8 is because of not only the limitation of the dynamic range of the hologram and the digitization but also other distortions which arise during the recording and reconstruction.

Fig. 4.11

time that the holograms are being recorded, without any expenditure of processor time [34].

One way to form a spatial carrier, as described in Section 3, is to set the period of the spatial carrier equal to two readings of the mathematical hologram and to set the digitization interval during the hologram recording equal to four times the period of the spatial carrier. In practice, this procedure means that in the course of the recording the sign of all the odd pairs of rows of the mathematical hologram should be changed (each pair of rows on the mathematical hologram is a row of real parts and a row of imaginary parts). The result of introducing such a spatial carrier can be seen from Figs. 4.14 and 4.15, which are images formed from holograms with and without the introduction of a spatial carrier, respectively. In Fig. 4.15 there is a superposition of the images formed from the even and odd parts of the hologram {any function $f(x)$ can be written as the sum of an even function $\frac{1}{2}[f(x) + f(-x)]$ and an odd function $\frac{1}{2}[f(x) -$

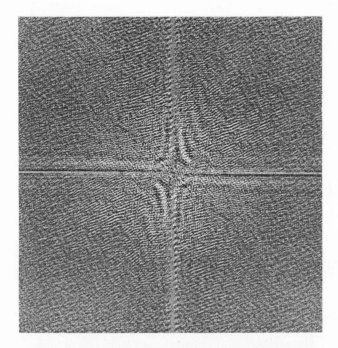

Fig. 4.12

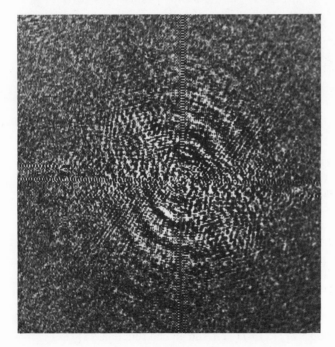

Fig. 4.13

Fig. 4.14

Fig. 4.15

$f(-x)]$}. The spatial carrier shifts these images by an amount precisely equal to half the distance between the diffraction orders corresponding to the raster of the recorded hologram. From the reconstruction standpoint this method for introducing a spatial carrier is equivalent to making the object symmetric by doubling.

If yet another spatial carrier is introduced, along a different direction (along the rows of the mathematical hologram), and if the period of this carrier is equal to one reading of the hologram and two digitization intervals of the recording device, then the image which is formed is displaced by half a diffraction order in this same direction (Fig. 4.16a). This procedure is analogous to a quadrupling of the object. Shown for comparison in Fig. 4.16b is the result of a reconstruction from a hologram obtained by quadrupling the object with mirrors.

The period of the spatial carrier can be doubled in one direction—along the row of the mathematical-hologram matrix. This approach is a version of Lee's method [27] in which the redundancy of the physical hologram is used in a more compact way. In a recording by the Lee method, the positive and negative values of the real and imaginary parts of the reading of the mathematical hologram are arranged in a linear fashion in one row of the physical hologram so that the recorded hologram does not have the same digitization step along two directions. In practice this circumstance means that the resolution of the recording device is not being exploited fully. In our method, the values of the readings are arranged in a "quartet" in a plane so that the asymmetry of Lee's method is eliminated. Figure 4.17 shows the result of reconstruction when a spatial carrier is introduced. The carrier has the implication that the images in one direction are shifted by half the diffraction order on the raster, while in the other direction they are shifted by an entire order.

The spatial carrier is not simply a method for recording complex-valued holograms, but is also a tool for transforming the holograms themselves. One useful property of this carrier is the image shift. It is possible to move the object without resynthesizing the hologram. This property may find applications in the synthesis of stereo mosaics (Section 5).

By introducing a spatial carrier with a period which is a linear function of the reading index, we can confer on the synthesized Fourier hologram the properties of a lens, essentially converting it into a Fresnel hologram (Section 1). If a spatial carrier of this type is combined with the carrier used to record the complex-valued holograms, it is possible to focus the two conjugate images, real and imaginary, in different planes. This circumstance is illustrated by Figs. 4.18a and 4.18b, which are images formed from such a hologram in its two focal planes: the front focal plane (a), in which a direct image is formed, and the rear focal plane (b), where the conjugate image is reconstructed. In the Fourier plane (c) both of the reconstructed images (direct and conjugate) are greatly out

Fig. 4.16

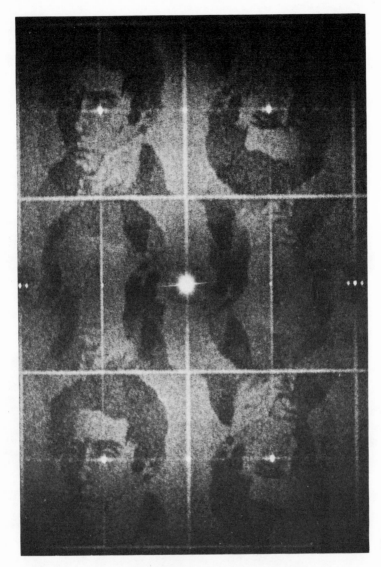

Fig. 4.17

of focus. This effect can be utilized to produce holograms which form an image in different planes. It is sufficient to introduce a "Fresnel" carrier separately in the holograms of the different planes and then sum the resulting holograms. Some similar methods for synthesizing holograms of multiplane objects are described in [58]. The result of this summation of Fourier holograms without the "Fresnel" carrier and of a hologram with a carrier is illustrated by Fig. 4.19,

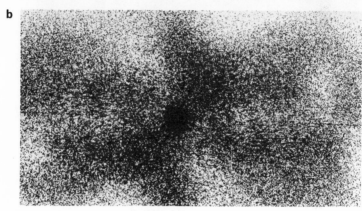

Fig. 4.18

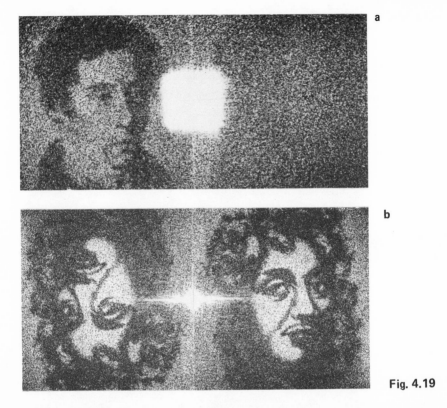

Fig. 4.19

which is an image formed from the resultant hologram in the Fourier plane. The image in Fig. 4.19a is reconstructed in the Fresnel plane of such a hologram.

Hologram "Multiplexing" as a Method for Matching the Recording Medium. As was already mentioned, the problem of matching the dynamic range of the hologram to the dynamic range and amplitude resolution of the apparatus and media used in recording holograms is one of the key problems of holography, which ultimately determines whether the holographic techniques can be used in various scientific and engineering fields. The most common matching method, in both physical and digital holography, has been to use a random diffuser, but the introduction of a diffuser in the hologram-synthesis arrangement leads to an additional speckle noise on the resulting images. It has been suggested [35-39] that a "deterministic diffuser" be used; such a diffuser would "smear" the information over the hologram area in precisely the same way as a random diffuser, but it would not produce a noise pattern on the images. It may be that this concept of a deterministic diffuser will be implemented most extensively in digital holography since in this case the problem of devising the appropriate physical device will not arise. One way to introduce a regular redundancy in a digital hologram is to use "multiplexing" of a nondiffuse calculated hologram [40] : A Fourier hologram stored in a computer is broken up into several parts with different signal intensities, as shown in Fig. 4.20, where I is the signal intensity and f is the spatial frequency. Then the signal from the central region, which corresponds to the lowest spatial frequencies of the image (which are usually the most intense; region ab on curve 1 in Fig. 4.20), is attenuated at each point by a factor of L. Here L is on the order of the ratio of the maximum signal in this region to the maximum signal in an adjacent, less intense region. After the attenuation by a factor of L, the central region is repeated $m \times n = L$ times over the area of the hologram. Upon superposition, the signals are added. This procedure can be repeated several times, as shown in Fig. 4.20 (curves 2 and 3). After repeated duplication and superposition, we have a multiplex hologram, and it is this hologram which is recorded.

We turn now to some results on the multiplexing of a nondiffuse hologram, carried out in accordance with

$$\tilde{B}_{kl} = B_{kl} + L^{-1}B_{i+i_0,\, j+j_0},\qquad (4.1)$$

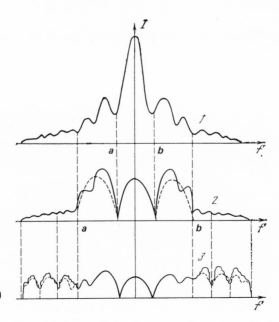

Fig. 4.20

where $\{B_{i+i_0,\ j+j_0}\}$ is the array of complex numbers of the region of the hologram being duplicated, $\{B_{kl}\}$ is the array of complex numbers of that part of the hologram in which the duplication is carried out, $\{\tilde{B}_{kl}\}$ is the array of complex numbers after the multiplexing, L is the factor by which the hologram elements of the multiplexed region are attenuated, i and j are the coordinates on the multiplexed fragment (the index of the row and the index of the element, respectively), i_0 and j_0 are the initial values of these coordinates, and k and l are the coordinates of the hologram element in the multiplex region, given by

$$k = 512 - 2\ i_{\max} + mi,$$
$$l = 512 - 2j_{\max} + nj.$$

Here m and n are the numbers of the fragments being multiplexed; $mi = 0, 1, 2, \ldots, mi_{\max} - 1; nj = 0, 1, 2, \ldots, nj_{\max} - 1; i, j = 0, 1, 2, \ldots, 255$.

The number of multiplexed fragments is varied from one to four, while the attenuation factor is chosen empirically on the basis of a digital printout of the calculated Fourier hologram. This factor is approximately the ratio of the maximum value on the multiplexed fragment to the maximum value on the adjacent fragment (for the hologram under study, $L = 16$). The initial coordinates of the first multiplexed fragment are $i_0 = j_0 = 504$; the dimensions of this fragment are 8×8 elements. In a repeated multiplexing, i_0 and j_0 as well as the dimension of the multiplexed fragment are changed in a programmed manner. The last multiplexed fragment has dimensions of 256×256 elements.

Figure 4.21 is the image formed from the hologram in Fig. 4.22 after a fourfold multiplexing of the regions of the nondiffuse hologram in accordance with (4.1). This image shows some loss of sharpness as well a rectangular grid structure. The loss of sharpness can be attributed to the attenuation of the higher spatial frequencies on the hologram in the course of the repeated duplication of the intermediate and low spatial frequencies, as a result of which the image becomes "smoother." The rectangular grid is a consequence of the duplication of the spectral components. The grid frequency depends on the dimension of the multiplexed fragments. This grid structure degrades the image quality, so that an effort was made to determine the effect of the original dimension of the multiplexed fragment upon its duplication from different parts of the hologram. Figure 4.23 is the image formed from the same hologram with $i_0 = j_0 = 504$; the dimensions of the multiplexed fragment are 8×8 elements, and $L = 16$. The dimensions of the fragment $\{B_{kl}\}$ in which the multiplexing is carried out are 64×64 elements. This image is rich in halftones, so it is better than that in Fig. 4.24, which was formed from an ordinary nondiffuse hologram which had not been subjected to multiplexing.

These results demonstrate the potential of this method for narrowing the dynamic range of the hologram. On the one hand, this method makes it possible to offset the effects of the coarse digitization of the peripheral parts of the holo-gram (this digitization leads to an image consisting of bare outlines); on the other hand, it makes it possible to avoid the additional speckle noise which arises when a random diffuser is incorporated in the hologram-synthesis arrangement.

Experiments with a "Spoiled" Fourier Transformation. The computer time required per hologram is one of the primary factors governing the feasibility of using such holograms in applications. The invention of the FFT algorithm has made the computer synthesis of holograms practicable. The number of opera-tions required to perform the Fourier transformation (the most laborious operation in hologram synthesis) can be reduced further by simplifying the operations of multiplying by a complex exponential function, which are carried out in the course of the transformation.

In this subsection we discuss a study of the possibility of accelerating the multiplication through a coarse digitization of the values of the real and imagi-nary parts of the complex exponential function which is the kernel of the Fourier transformation, and we discuss the effect of this digitization on the result of the reconstruction from the synthesized hologram.

This procedure for simplifying the hologram synthesis can be justified by two approaches. We will discuss these approaches for the case of the digitization of sines and cosines at three levels: $-1, 0, 1$. In this case it is completely unnecessary to carry out the multiplication operation in the course of the Fourier transformation, and some of the addition operations also can be eliminated.

First approach. The coarse digitization of the real and imaginary parts of the complex exponential function, in three levels, can be thought of as the result of using a "pseudodiffuser" at the object which changes the phase of the exponen-tial function in a definite manner to the nearest of the values $-\pi, 0, \pi$. Since a pseudorandom diffuser is attributed to the object in the hologram synthesis, there is reason to hope that this diffuser will randomize the effect of the deter-minate diffuser and that it will not severely affect the reconstruction from the hologram synthesized in this manner.

Second approach. The result of the coarse digitization at three levels is the conversion of a sinusoidal signal into rectangular pulses. These pulses can be expanded in Fourier series; the first harmonics of the series give the ordinary exact Fourier transform, while the others also give a Fourier transform but with an enlargement of the frequency scale by a factor equal to the index of the har-monic. The holograms corresponding to these "harmonic" transforms, when superimposed on the "fundamental" hologram, produce troublesome distorted images. The intensity of these troublesome components can be reduced by appropriately choosing the point for the switch from 0 to 1 and from 0 to -1. For example, it is simple to show that if the first harmonic is required to have the maximum power, the ratio of the interval of the argument of the sinusoidal

Fig. 4.21

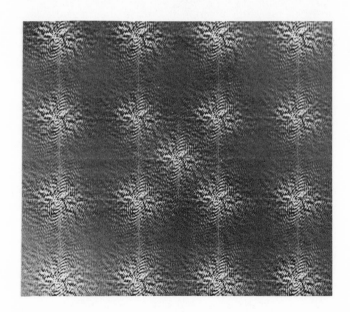

Fig. 4.22

Fig. 4.23

Fig. 4.24

function between the point at which this function crosses 0 to the point of the switching to ±1, on the one hand, to the period of the sinusoidal function, on the other hand, should be about 0.37.

Figure 4.25 shows the result of experiments with holograms synthesized with a "spoiled" Fourier transformation. This image was formed from a hologram calculated in the digitization of the values of an exponential function at five levels in accordance with

$$\sin \varphi = \begin{cases} 0 \ , & 0 \leqslant \varphi < \pi/16, \\ 1/2, & \pi/16 \leqslant \varphi < \pi/8, \\ 1 \ , & \pi/8 \leqslant \varphi \leqslant \pi/2. \end{cases} \tag{4.2}$$

This image is essentially the same as those formed from ordinary synthesized holograms (e.g., Fig. 4.8).

In the synthesis of diffuse holograms, the severe distortion of the Fourier transform resulting from the digitization of the values of the complex exponen-

Fig. 4.25

tial function thus does not destroy the image, although there is a definite effect on the image quality. The effect on the image quality also depends on the FFT code used. For different transformation procedures, the digitization can apparently have different effects on the transformation accuracy.

Forming Images from Computer-Synthesized Holograms. In experiments with synthesized holograms, visual observation or photography is used in some stage of the image simulation to obtain qualitative and quantitative estimates of the calculated results. In the course of the recording, both the noise produced by the hologram itself (the speckle, the digitization, the effects of the limited dynamic range, etc.) and the noise and distortions which arise in the course of the reconstruction and photography are superimposed on the image. In this subsection we will describe some methods for combating these distortions.

A characteristic feature of the images formed from synthesized holograms (both diffuse and nondiffuse) is noise near the zeroth diffraction order. This noise results from the amplitude and phase distortions produced by the photographic materials used to record the hologram and the image formed from the hologram. The amplitude noise, which is primarily because of the graininess of the photographic emulsion, causes a random scattering of the laser wave used in the reconstruction and of the reconstructed wavefront. The phase noise, on the other hand, is because of the optical inhomogeneity and deformation of the surfaces of the hologram and of the photographic film.

There are several ways to combat these types of noise. First, the image can be shifted away from the central spot by introducing a spatial carrier or by placing the edge of the original image on part of the raster used. In either of these cases, the shift of the resulting image with respect to the zeroth diffraction order and the shift of the noise are achieved at the cost of using some of the readings (the elements of the image and the hologram) to transmit the spatial carrier. A second way to combat the noise is to use a superposition method: Several copies of the hologram are prepared and arranged in a mosaic. The reconstruction is carried out from this mosaic hologram. Figure 4.26 shows an image formed from a mosaic hologram consisting of nine elementary holograms each containing 1024×1024 elements. The dimensions of this hologram, recorded with a 25-μm raster, are 7.6×7.6 cm. The quality of this image is essentially as good as that of the original image.

Another effective way to combat the phase noise of the photographic film is to use liquid immersion compensators, i.e., a glass cell with optically flat outer walls which is filled with an immersion liquid, in which the synthesized hologram is placed. A special holder is used to position the cell in the reconstruction arrangement. The immersion cancels the fluctuations in the film thickness because of both irregularities of the substrate and fluctuations in the silver density on the developed hologram. The light scattering is accordingly reduced, especially at the center of the image, and the overall image contrast

Fig. 4.26

Fig. 4.27

is improved. The immersion liquid used in these particular experiments was cedarwood oil, whose refractive index, $n = 1.51$ is approximately equal to both the refractive index of the emulsion and the substrate ($n = 1.50$), on the one hand, and that of the glass cell ($n = 1.52$), on the other.

The image noise can be reduced further by placing an apodizing mask on the hologram. Figure 4.27 shows the image from a mosaic of 16 holograms, each containing 1024×1024 readings, for the case in which an apodizing mask of the type

$$\varkappa = 1 - [x^2/X^2_{\max} + y^2/Y^2_{\max}]/2, \tag{4.3}$$

is placed on the mosaic. Here x and y are the coordinates on the hologram, and $X_{\max}$ and $Y_{\max}$ are the coordinates of the extreme points. The apodizing mask described by (4.3) smooths the interpolating function which converts the readings into a continuous image (Section 3), and it suppresses the interference between adjacent readings which occurs if the hologram is illuminated through a square window.

Fig. 4.28

To conclude this subsection we note that the images formed from the synthesized holograms in our experiments were recorded on type NR-20 photographic plates, high-sensitivity facsimile paper, and photographic film of several types (KN-1, KN-2, KN-3, A-2, and Mikrat-300). Despite the advantages of the photographic plates and the photographic paper, the type KN photographic films proved most convenient. The chemical development was carried out in the solutions recommended by the manufacturer. The development was controlled with a standard sensitometer wedge, which was placed on the hologram in a particular way and obtained along with the image during the reconstruction.

Synthesis of the Elements of Optical Signal-Processing Systems. The methods described above for synthesizing and recording holograms can be used for computer synthesis of the elements of optical systems (lenses, diffraction gratings, etc.) and of complicated optical complex-valued filters for optical signal-processing systems. We will discuss some examples of the synthesis of such elements and filters. Figure 4.28 is a photograph of a positive lens which was synthesized on a computer and recorded with a spatial carrier (Section 3). The

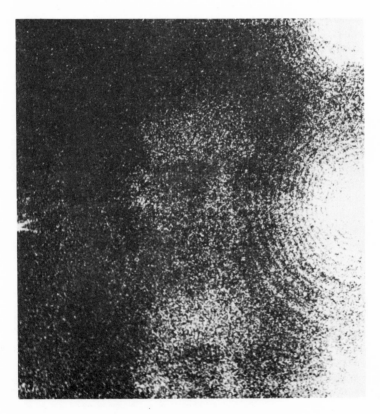

Fig. 4.29

time required to design this lens, consisting of 1024 × 1024 elements, is 5 min. Figure 4.29 is an image formed from a hologram by means of the computer-synthesized lens shown in Fig. 4.28; this lens performs the Fourier transformation in the reconstruction arrangement. Figures 4.30a and 4.30b show some synthesized spatial filters for reconstructing defocussed images; Fig. 4.30a shows a "pseudoinverse" filter recorded by a modification of Lee's method, and Fig. 4.30b shows the amplitude part of a Wiener filter (see [59], for example, on the use of synthesized filters for image reconstruction).

Section 5. Experiments on the Visualization of Three-Dimensional Objects

Synthesis of Stereo Holograms and Stereo Mosaics. The synthesized Fourier holograms can reproduce for visual observation a projection of an object on a plane perpendicular to the observation direction. These holograms can convey spatial relations only to the extent that these relations affect the projections, i.e., through perspective distortions and through the masking of some parts of the object by others. Fourier holograms of objects cannot convey the depth of various planes within the object. If a spatial carrier which simulates a lens is incorporated in the Fourier holograms, it is possible to synthesize holograms which focus different objects or different cross-sectional views of the same object in different planes and which carry out the reconstruction in these planes (Section 4). It is still not possible, however, to convey the three-dimensional nature of objects in a convenient and natural way for visual observation. Because of our binocular vision, an important factor in the perception of three dimensions during observation under natural conditions is the stereo effect: the subjective perception that an object is three-dimensional if it is observed with the two eyes from different perspectives or with parallax. This effect can be reproduced with digital holograms if separate holograms are synthesized for different perspectives of the object and then observed simultaneously by two eyes. We will refer to such holograms as stereo holograms [40, 60].

To check the possibility of observing the stereoscopic effect with computer-synthesized Fourier holograms, we carried out a series of experiments with holograms of test objects having various dimensions and with various degrees of parallax [41, 42, 61-63]. These test objects were synthesized by a computer; their geometric dimensions and coordinates on the image could be changed through programming. In this manner we produced the original stereo pairs of transparencies, which were then used to synthesize the holograms intended for the right and left eyes. To increase the area and for convenience in observation, the synthesized holograms were duplicated and arranged in a mosaic. These mosaic holograms were squeezed between two optically flat glass plates, whose inner surfaces were coated with immersion fluid. The glass plates with the holo-

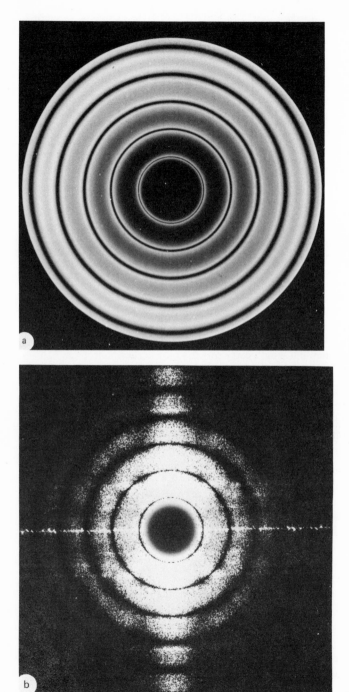

Fig. 4.30

grams were then placed in an eyeglass frame. When an observer puts these "eyeglasses" on and looks through them at a point source of monochromatic light, he or she can observe the stereoscopic effect at the test object.

Figure 5.1 shows images formed from stereo holograms. It can be seen that when the observer looks at the digital stereo hologram he or she simultaneously sees two conjugate images with the same eye and in the same plane. If the stereoscopic effect is observed at one image (if one test object is closer to the observer), then a "pseudoscopic" effect is observed at the second image (the object which had previously been in front is now to the rear). For the formation of the images shown in Figs. 5.1a and 5.1b we thus observe five vertical planes: the central plane which coincides with the plane of the point source, two planes in front of this source, and two behind it. The experiments showed that the magnitude of the parallax, i.e., the shift of the corresponding image points, affected the observer's perception of the stereoscopic effect when looking at the stereo holograms. Specifically, in the case of a large parallax the stereoscopic effect was perceived after a certain delay, rather than instantaneously. The larger the parallax, of course, the greater the subjectively perceived depth of the planes. By incorporating in the program a change in the parallax of the objects specified numerically in the computer, it is possible to synthesize holograms which create the impression of a change in depth of the planes.

If several holograms are synthesized for several slightly different values of the object parallax, a mosaic or composite stereo hologram can be produced for each eye. Figure 5.2 shows such a mosaic, consisting of three holograms. By looking through different parts of this composite hologram one can observe the test object in various planes. If the eye is moved continuously along the mosaic, one can observe a continuous transition from one plane to another.

The successful experiments with stereo holograms made it possible to use the method of synthesizing stereo holograms to produce a three-dimensional holographic film which demonstrates the rotation of a three-dimensional object

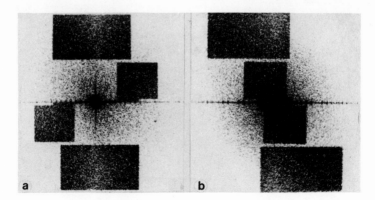

Fig. 5.1 a b

Fig. 5.2

[43]. For convenience and simplicity in the programming and recording of
the synthesized holograms, the object of the visualization was chosen to be a
symmetric object: two spheres connected by a cylinder (a "dumbbell") rotating
around an axis passing through the center of the cylinder. The projections of this
object onto a plane perpendicular to the observation direction consist of two
circles connected by a rectangle, for all perspectives. As this object is rotated,
the projected distance between the centers of the circles changes. The rotation
of the dumbbell was simulated by calculating 18 projections onto planes. The
projections corresponded to 18 perspectives, covering an angle of 180° at 10°
intervals. The other perspectives required for the full 360° rotation did not have
to be calculated because of the symmetry of the object. Diffuse Fourier holograms were synthesized for each perspective, and then 20 duplicates were made

Fig. 5.3

of each hologram. The resulting 360 holograms were arranged in the corresponding order along a closed ring 50 cm in diameter (Fig. 5.3), thereby forming a
mosaic (composite) hologram *1* which was 37 mm high and about 157 cm long.
This hologram was held in an annular holder *2* which was mounted on the shaft
of an electric motor. The hologram was illuminated with a spherical laser beam†
with the help of a reflecting cone, *3*. The holograms are arranged in the ring in
such a manner that, when the observer looks through them with both eyes at
the light source, different holograms, calculated for adjacent perspectives, will
be in front of the two eyes. The fixed holograms thus create the visual perception that the object (the dumbbell) is three dimensional. One of the spheres
appears closer than the source to the observer, while the other appears to be
behind the source. As the observer moves along the holograms he or she sees
the dumbbell rotate. If several people observe the hologram simultaneously,
each sees a dumbbell suspended in space at the corresponding perspective. When
the motor is turned on and the hologram starts rotating, there is a clear impression of a smooth rotation of the dumbbell in a certain direction (clockwise
looking down). The continuous rotation of the object, without hesitation or
jumps, allows us to speak in terms of a motion picture effect as the observer
looks at the rotating composite hologram. Each hologram produces an image of
a moving object in some static phase of its motion. Several frames showing the
different phases of the motion to the observer in a certain sequence and at a
certain frequency (sufficient to make the motion appear continuous) completely
simulate the motion of the object, creating the illusion of rotation. In the
present experiments, the motor was run at 12 rpm, so that 7.2 perspectives
passed the observer per second. Even at a much lower velocity, however, the
observer still sees a continuous change.

Another unique feature of the digital holographic film is that the frames
(or Fourier holograms) can be drawn continuously since the Fourier transformation is invariant with respect to displacement. This feature represents an important advantage of holographic film over ordinary film, which requires various
types of shutters and complicated mechanical devices for advancing the motion
picture film.

The method described here for synthesizing stereo holograms can be used to
visualize objects which are moving at different velocities. That this effect can
be observed is confirmed by an examination of a synthesized holographic film
which demonstrates the rotation of two three-dimensional objects (again,
dumbbells) at different rates. In this case it is necessary to calculate 36 projections of the dumbbells onto planes in order to simulate the rotation. These projections correspond to 36 perspectives, again at intervals of $10°$, which span an

---

†It is not necessary to use a laser. Good results can be obtained with the incoherent light
from an incandescent lamp with a filter, e.g., a ZS-2 filter.

angle of 360° (a complete revolution) for one of the dumbbells and an angle of 180° for the other (half a revolution). Three of these projections of the object are shown in Figs. 5.4a–5.4c. When the motor operates at 12 rpm the dumbbells rotate smoothly, and the perspectives pass the observer at rates of 7.2 and 3.6 perspectives/sec, respectively. The total number of holograms in the film is 720; the length of the mosaic is 172 cm.

As mentioned earlier, if the stereo holograms are Fourier holograms, the observer sees two conjugate images. In principle, this is inconvenient; but if a spatial carrier simulating a lens is used (Sections 2 and 4), it is possible to focus the conjugate images in different planes so that the observer will not see one of these images. The kinoform method also holds promise here [52]. Karnaukhov *et al.* [64] have recently reported the synthesis of a kinoform film, consisting of 1152 elementary kinoforms, which demonstrates the rotation of an asymmetric object.

The stereoscopic effect of the synthesized holograms opens up real possibilities for using digital holography in the visualization of three-dimensional objects which are specified by a mathematical description. This stereoscopic effect also holds definite promise for the development of holographic three-dimensional television. In holographic television, the stereoscopic holograms could be synthesized at the receiver from a video signal of images of various perspectives of the scene. A television system of this sort, with hologram synthesis at the receiving end, is convenient for using the methods of intraframe transformation image coding to reduce the redundancy.

Finally, we would like to point out another approach which could substantially reduce the computer time required for synthesizing stereoscopic holograms of objects consisting of several movable planes. In this approach, the theorem on the displacement of discrete Fourier transforms (Appendix II) is used to synthesize a set of stereoscopic holograms of opaque objects with various degrees of parallax from several holograms, synthesized for each plane of the object. The holograms for the various planes would not have to be constructed from scratch; it would be sufficient to multiply the original holograms

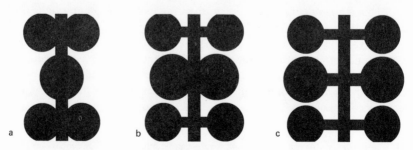

Fig. 5.4

by a complex exponential function, with a spatial frequency corresponding to the desired parallax. The saving in computer time for each stereoscopic hologram is by a factor of about $\log_2 N$, where N is the number of elements in the hologram. Since high-quality holograms require some 10^6 elements, the saving can be substantial.

Synthesis of Holograms with a Programmable Diffuser. Another important factor in the perception of three dimensions consists of the highlights at the concave and convex surfaces of the object. Highlights are important in the perception of three dimensions for objects which do not have well-defined contours and details useful for the stereoscopic effect and also in the perception of three dimensions with monocular vision, because of the play of highlights which occurs as the eye moves. This effect is used widely in pictorial art, and it contributes to the perception of three dimensions in photographs. Highlights occur when an object is illuminated with directed light, because diffusely reflecting objects do not scatter incident light isotropically. Because of this characteristic, the intensity of the light reflected by a certain part of the surface in a given direction depends on the angle between this direction and the normal to this part of the surface and also on the direction to the light source.

The highlight effect can be reproduced in the synthesis of Fourier holograms of objects which are specified numerically [40, 63]. In visualization problems, it is important to convey the brightnesses and macroscopic shape of the object (i.e., irregularities which are substantially larger than the wavelength of the light source). These irregularities are described by the distribution of the modulus of the reflection coefficient over the surface of the object, referred to the plane tangent to the object and normal to the observation direction, and by the geometric shape of the surface, from which it is possible to find the distance from each point on the surface to this plane along the observation direction. In order to convey the diffuseness of the surface, it is necessary to supplement the "regular" component of the phase of the reflection coefficient, which describes the shape of the surface, with a "random" component, which describes the microscopic shape of the surface and which simulates the diffuseness. In order to simulate the nonuniform scattering of light in different directions by a diffuse surface, this random phase component must be correlated: Its energy spectrum must be governed by the angular distribution of the reflection intensity of a diffuse surface at the given point. The uncorrelated component, or a diffuser with uncorrelated readings, corresponds to reflection of the incident light which is uniform in all directions.

Figure 5.5 shows a sequence of operations which can produce this random correlated phase component. Blocks *1–4* provide the two-dimensional algorithm for obtaining correlated Gaussian numbers; this algorithm is described for the one-dimensional case in [44]. In principle, the mask—the directional pattern of the plane diffuse surface in block *3*—should be chosen from the given angular

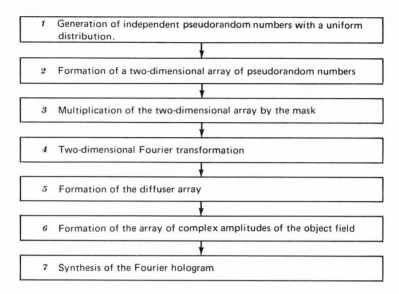

1	Generation of independent pseudorandom numbers with a uniform distribution.
2	Formation of a two-dimensional array of pseudorandom numbers
3	Multiplication of the two-dimensional array by the mask
4	Two-dimensional Fourier transformation
5	Formation of the diffuser array
6	Formation of the array of complex amplitudes of the object field
7	Synthesis of the Fourier hologram

Fig. 5.5

Fig. 5.6

distribution of the reflection intensity for the given diffuse surface. By this we mean that the energy spectrum of the array $\{\exp i\varphi(k, l)\}$ is specified at the output of block 5, and it must be used to find the spectrum of the array $\varphi(k, l)$: the mask $h(k, l)$. This conversion is possible in principle, but in practice it is not really necessary since the simulating directional pattern would need to correspond to the actual pattern only roughly for visual simulation. It is sufficient to choose as the mask $h(k, l)$ some simple function which approximates the necessary directional pattern at least satisfactorily (primarily in terms of the number and width of the modes or lobes); the scale for the numbers $\varphi(k, l)$ should be chosen such that the relative number of readings $\varphi(k, l)$ outside the interval $(0, 2\pi)$ is small. In this case the energy spectrum of the array $\{\exp i\varphi(k, l)\}$ differs negligibly from the spectrum of the array $\{\varphi(k, l)\}$ and thus from the desired directional pattern. It should also be noted that the array $\{\exp i\varphi(k, l)\}$ does not have to be resynthesized for each hologram; it can be a universal diffuser routine, used repeatedly to synthesize holograms of different objects which have identical or similar diffuse properties. In block 6, this diffuser is combined with the regular component of the phase of the object and with its intensity reflection coefficient to produce the input array for synthesizing the Fourier holograms.

We will refer to those holograms which are synthesized by this procedure as "holograms with a programmable diffuser." In principle, a programmable diffuser permits computer generation of Fourier holograms which contain information on all perspectives of the object and thus on the shape of the object. This conclusion is verified by experiments with such holograms [45].

In the experiments on synthesizing the universal diffuser routine, the three-dimensional form of the mask was a tetrahedral pyramid. Figure 5.6 shows the angular distribution of the light intensity for such a diffuser; this figure is the result of a reconstruction from a hologram of a diffuser with a random phase distribution, obtained after block 7 without the use of block 6 in Fig. 5.5. Plotted along the horizontal and vertical axes in Fig. 5.6 are the horizontal and vertical observation angles, respectively. We see that there is a preferred direction in which the surface reflection is more intense than in others.

This diffuser was used in the method described here to synthesize Fourier holograms for three versions of three-dimensional, uniformly colored surfaces: a corrugated surface (Fig. 5.7), a pyramid (5.8), and a hemisphere (Fig. 5.9). The illumination direction was assumed to be the vertical direction in Fig. 5.7-5.9. The holograms obtained for these objects are shown in Figs. 5.10-5.12, respectively. The hologram of the corrugated surface (Fig. 5.10) is particularly instructive. In a sense, it consists of two parts, each containing information on planes in one orientation. If we look through these holograms at the source of a spherical wave, moving the pupil of the eye over the hologram surface, we observe a displacement of the highlight, i.e., the brightest spot on the image.

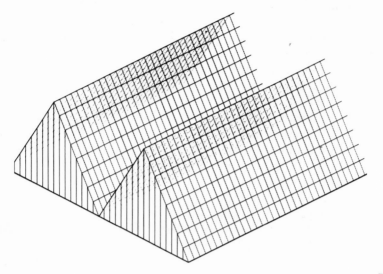

Fig. 5.7

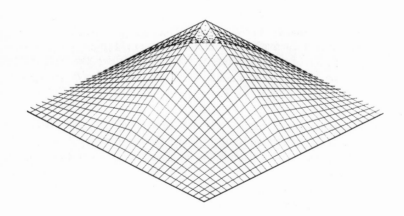

Fig. 5.8

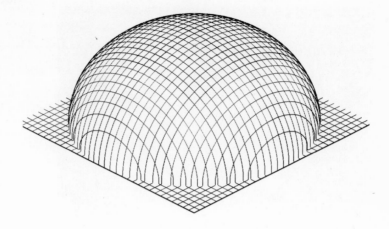

Fig. 5.9

Fig. 5.10

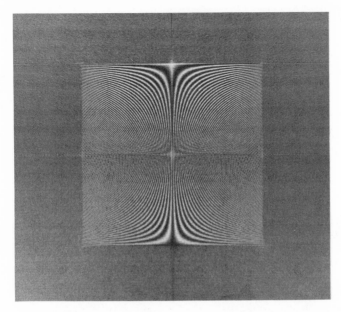

Fig. 5.11

Fig. 5.12

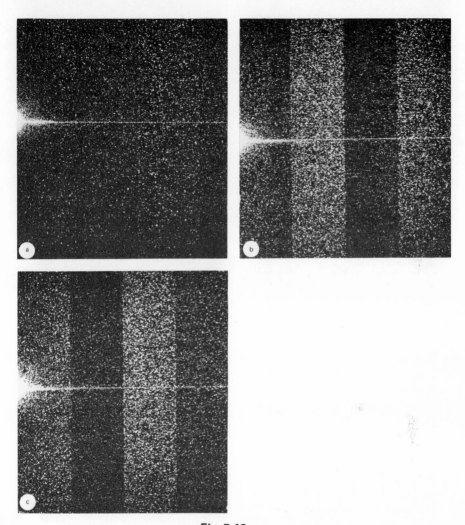

Fig. 5.13

The effect is precisely the same as if we were looking from different directions at the real objects (the corrugated surface, the pyramid, and the hemisphere), illuminated by a light beam. Figures 5.13–5.15 show images observed from different parts of these holograms.

Figure 5.13 shows images obtained from the hologram of the corrugated surface: from a part of the hologram near its center (a), at the right side (b), and at the left side (c). Figure 5.14 shows images obtained from the hologram of the pyramid: from the center of the hologram (a), from the upper left part of

Fig. 5.14

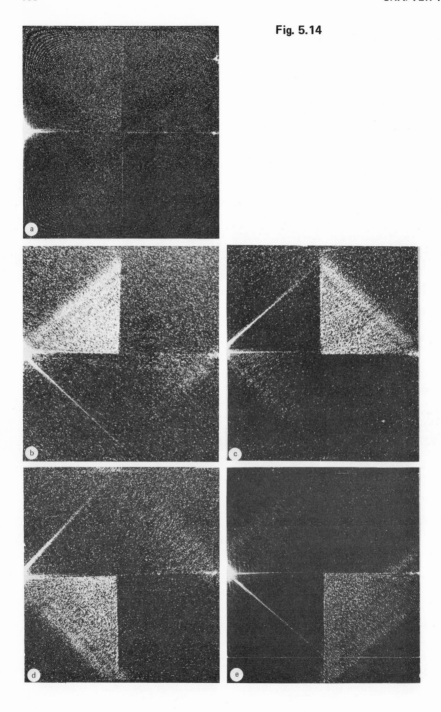

Fig. 5.15

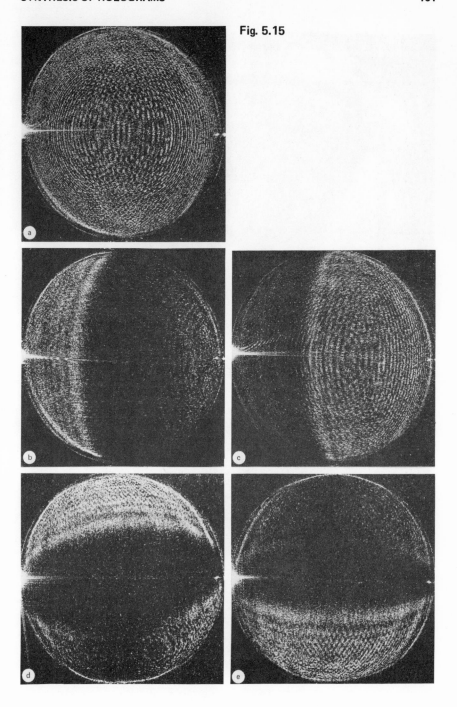

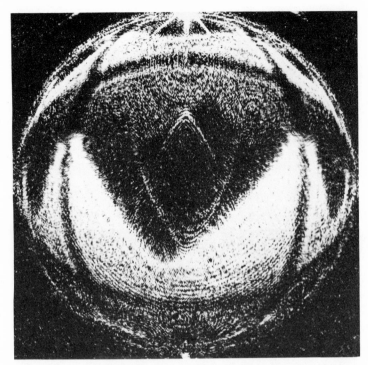

Fig. 5.16

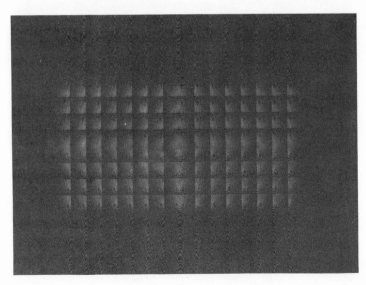

Fig. 5.17

the hologram (b), from the upper right (c), from the lower left (d), and from the lower right (e). Figure 5.15 shows images obtained from the hologram of the hemisphere: from the center of the hologram (a), from the left half (b), from the right half (c), from the upper half (d), and from the lower half (e). In other words, Figs. 5.13–5.15 are the images which would be observed from points in front of the object, to its left, to its right, above it, and below it.

Curiously, if the reconstruction from this hologram is carried out while a mask with some appropriate shape covers part of the hologram, the reconstruction can lead to very unusual effects which could not be produced under natural conditions. As an example, Fig. 5.16 shows the result of a reconstruction from the hologram of a sphere with a triangular mask. We see from these figures that a single hologram with a programmable diffuser contains information on all perspectives of the object in the solid angle ($\pm\pi/2$, $\pm\pi/2$).

The holograms used in these experiments have 1024 × 1024 elements and are recorded with a 25-μm raster. The physical size of the hologram is 25 × 25 mm. All perspectives of the object can be observed within windows 25 × 25 mm in size. The area of 25 × 25 mm is small in comparison with the size of the eye, and such holograms are not convenient for inspection. It is not possible to look at them with both eyes. It would be more convenient to enlarge the angular scale of the hologram, and this can be done by breaking the hologram up into parts, reconstructing the various perspectives from these parts, duplicating these parts, and arranging them in a mosaic. Figure 5.17 shows an example of this duplication for the hologram of Fig. 5.12. A hologram consisting of 1024 × 1024 elements has been broken up into 16 fragments each 256 × 256 elements in size. With recording by means of a 25-μm raster, each fragment is duplicated four times in the horizontal direction and eight times in the vertical direction. The result is a composite macrohologram with a programmable diffuser containing 4096 × 1892 elements.

2

Wavefront Analysis and Simulation

The analysis and reconstruction of holograms by computer and the digital simulation of holographic processes constitute a couple of the most important applications of computers in holography. The versatility of digital computers, the fact that they can be programmed, the accuracy and reproducibility of the results, the freedom to break into the program and alter it in any stage, and the ability to obtain essentially any quantitative results in any stage of the processing make the digital computer potentially one of the most convenient facilities for wavefront analysis and simulation.

We will restrict the present illustration of digital solution methods to the three applications of most practical interest: the formation of images of objects from their holograms; the determination of the directional pattern of an antenna from field measurements at the antenna aperture; and the simulation of diffuseness of the object and of distortions in the recording of the holograms on the result of the reconstruction. Another important practical problem—the reconstruction of interferograms with digital computers—is discussed in [46].

Section 6. Digital Reconstruction from Holograms

Mathematical Model and Discrete Representation. The physical hologram is the recorded result of the interference of the wavefront scattered or radiated by the object and a reference wave. The problem of forming an image of an object from its hologram is one of using the hologram to reconstruct the distribution of the wave intensity at the object. A related problem is that of determining the characteristics of radiating systems, in which case the results found from measurements of the amplitude and phase of the radiated wave on some surface must be used to determine the amplitude and phase distributions in some other spatial region.

The problem of forming an image from a hologram is one of the so-called inverse problems of mathematical physics. One way to solve such problems is to subject the result of the measurement—in this case, the hologram—to a transformation which is the inverse of that which relates the object and the result of the measurement; then, if necessary, the resulting solution can be corrected on the basis of *a priori* information on the object.

Let us consider the simplest and most frequently encountered cases: those in which the holograms have been recorded in the far zone (the Fraunhofer zone) or in the Fresnel zone, i.e., the cases in which the complex wave amplitude in the plane of the hologram is related to the complex field amplitude at the object by a Fourier or Fresnel transformation.

Let $\Gamma(\xi, \eta)$ be a complex function which describes the result of the recording of the wavefront by the hologram. This function may represent either the amplitude transmission of the optical hologram recorded on a photographic carrier or the result of a measurement of those components of an rf field or a sound wave which are in phase and orthogonal to the reference signal. In the case of far-zone hologram recording, the distribution of the complex field amplitude $b(x, y)$ at the object can be found by taking the inverse Fourier transform of the function $\Gamma(\xi, \eta)$:

$$ b(x, y) = \iint\limits_{(\xi, \eta)} \Gamma(\xi, \eta) \exp\left(i2\pi \frac{1}{\lambda d}(x\xi + y\eta)\right) d\xi \, d\eta, \qquad (6.1) $$

where λ is the wavelength and d is the distance from the source to the recording plane. In this case, these quantities determine the scale of the function $b(x, y)$ in the (x, y) coordinate plane.

The problem of determining $b(x, y)$ from (6.1) with a digital processor is related to the problem of using digital computers to synthesize Fourier holograms, discussed in Section 1. Using the results of that section, we can show that with a finite size of the hologram $\Gamma(\xi, \eta)$ the object is characterized by readings of the complex field amplitude $b(k, l)$, which are related to the hologram readings $\Gamma(r, s)$ by a DFT:

$$ b(k, l) = \sum_{r=0}^{N_x} \sum_{s=0}^{N_y} \Gamma(r, s) \exp\left[i2\pi\left(\frac{kr}{N_x} + \frac{ls}{N_y}\right)\right], $$

$$ k = 0, 1, \ldots, N_x - 1; \quad l = 0, 1, \ldots, N_y - 1, \qquad (6.2) $$

where the digitization step of the hologram is governed by both the necessary angular dimensions of the object, θ_x and θ_y, and the angle of the reference beam or the phase shift of the reference signal. If

$$2\theta_x = 2X_{\max}/d, \quad 2\theta_y = 2Y_{\max}/d \qquad (6.3)$$

are the angular dimensions of an object with linear dimensions $2X_{\max}$ and $2Y_{\max}$ along the x and y axes, observed at a distance d, and if α is the angle between the reference beam, which is normal to the y axis and the normal to the (ξ, η) plane,† then the following inequalities must be satisfied in order to prevent a self-superposition of the edge regions of $b(k, l)$ as the result of the repetitions caused by the digitization of the hologram:

$$\Delta\xi \leqslant \lambda/(2\theta_x + \alpha), \quad \Delta\eta \leqslant \lambda/2\theta_y. \qquad (6.4)$$

Frequently, however, the angular dimensions of the object and the angle of the reference beam are not known. In such a case, the digitization parameters of the hologram should be chosen on the basis of a condition equivalent to (6.4): The digitization steps along ξ and η must be such that one period of the maximum spatial frequency of the hologram corresponds to at least two readings of this hologram.

The results of measurements by devices which measure the field (field pickups) are usually not the actual readings of the continuous hologram,

$$\Gamma(r, s) = \iint\limits_{(\xi,\eta)} \Gamma(\xi, \eta)\, \delta(\xi - r\Delta\xi, \eta - s\Delta\eta)\, d\xi\, d\eta \;, \qquad (6.5)$$

i.e., the values of the hologram at the reading points $(r\Delta\xi, s\Delta\eta)$ as they would be singled out by a δ function; instead they are the values resulting from an averaging over the finite aperture of the detector,

$$\hat{\Gamma}(r, s) = \iint\limits_{(\xi,\eta)} \Gamma(\xi, \eta)\, H(\xi - \Delta\xi, \eta - s\Delta\eta)\, d\xi\, d\eta. \qquad (6.6)$$

As a result (as was shown in the analysis of analog media for recording holograms; Section 3), the readings $\hat{b}(k, l)$, which are calculated from $\hat{\Gamma}(r, s)$, differ from the true readings because of a masking

$$\hat{b}(k, l) = b(k, l)\, h(k, l) \;, \qquad (6.7)$$

by the function $h(k, l)$, which consists of the readings of the Fourier transform of the function $H(\xi, \eta)$,

---

†This condition is essentially a requirement on the arrangement of the digitization axes of the hologram. Specifically, these axes must be chosen such that one is normal to the direction of the reference beam or, equivalently, such that the direction of the other is the direction of the change in the reference signal.

$$h(x, y) = \iint\limits_{(\xi,\eta)} H(\xi, \eta) \exp\left\{i\frac{2\pi}{\lambda d}(x\xi + y\eta)\right\} d\xi\, d\eta, \qquad (6.8)$$

taken at the same points $(k\Delta x, l\Delta y)$, as the readings $b(x, y)$. To a certain extent, this masking can be offset if we know the aperture function of the measuring device, $H(\xi, \eta)$, and if we divide the values of $b(k, l)$ found from $\hat{\Gamma}(\xi, \eta)$ by $h(k, l)$ at those points at which $|h(k, l)| \neq 0$:

$$b(k, l) = \hat{b}(k, l) / h(k, l). \qquad (6.9)$$

A correction of this type cannot be carried out near those points at which $|h(k, l)|$ is approximately zero.

The problem of obtaining images from Fresnel holograms or calculating the field in the Fresnel zone is related in principle to the problem of synthesizing Fresnel holograms. There are certain differences in the approach to the choice of the digitization parameters of the holograms and the field to be calculated. Let us assume that $\Gamma(\xi, \eta)$ is a complex function which describes the recorded Fresnel hologram. Then the field amplitude at the object in the Fresnel zone is

$$b(x, y) = \exp\left[i\frac{\pi}{\lambda d}(\xi^2 + \eta^2)\right] \iint\limits_{(\xi,\eta)} \Gamma(\xi, \eta) \times$$

$$\times \exp\left[i\frac{\pi}{\lambda d}(\xi^2 + \eta^2)\right] \exp\left[-i\frac{2\pi}{\lambda d}(x\xi + y\eta)\right] d\xi\, d\eta, \qquad (6.10)$$

where λ is the wavelength and d is the distance at which we are seeking the field $b(x, y)$. In contrast with Fourier holograms, the parameters λ and d determine not only the scale but also the spatial distribution of the amplitude and phase of $b(x, y)$.

For analyzing Fresnel holograms (for reconstruction from them), we should distinguish between two situations: that in which it is necessary to reconstruct only the values of the light intensity at the object (the problem of forming an image of the object) and that in which it is necessary to determine both the amplitude and phase of the wave at the object. The first case is simpler. In this case we can omit the multiplication of the integral in (6.10) by the phase factor, and, in transforming to the discrete representation of the integral, we can choose the digitization step along x and y on the sole basis of the hologram dimensions (L_{hx}, L_{hy}):

$$\Delta x = \lambda d/L_{hx}, \qquad \Delta y = \lambda d/L_{hy}. \qquad (6.11)$$

With regard to the digitization along ξ and η, we note that the digitization depth and thus the dimensions of the aperture ot the measurement device which performs the digitization must be chosen to satisfy two conditions simultaneously. First, the discrete representation of the hologram must be sufficiently fine; in other words, the period of the maximum spatial frequency of the hologram must correspond to at least two readings [see condition (6.4)] . Second, the phase factor exp $[i(\pi/\lambda d)(\xi^2 + \eta^2)]$ in the integral in (6.10) must be conveyed accurately. The least stringent requirement which can be imposed here is that the digitization not disrupt the monotonic variation in the phase of the phase factor. This requirement means that the period of the maximum spatial frequencies of the exponential factor, which are obviously

$$\frac{d}{d\xi}\left(\frac{\xi^2}{2\lambda d}\right)\bigg|_{\xi\max} = \frac{\xi\max}{\lambda d} = \frac{L_{hx}}{2\lambda d}, \quad \frac{d}{d\eta}\left(\frac{\eta^2}{2\lambda d}\right)\bigg|_{\eta\max} =$$

$$= \frac{\eta\max}{\lambda d} = \frac{L_{hy}}{2\lambda d}, \tag{6.12}$$

must also correspond to at least two readings:

$$\Delta\xi \leqslant \lambda d/L_{hx}, \quad \Delta\eta \leqslant \lambda d/L_{hy}. \tag{6.13}$$

From these two pairs of estimates of $\Delta\xi$ and $\Delta\eta$, we must choose the smallest values for $\Delta\xi$ and $\Delta\eta$.

If we are also required to reconstruct the phase of the field $b(x, y)$, this dual approach must be taken in choosing the digitization steps along x and y. In other words, (6.11) must be supplemented with two other conditions which are analogous to (6.13):

$$\Delta x \leqslant \lambda d/L_{0x}; \quad \Delta y \leqslant \lambda d/L_{0y}, \tag{6.14}$$

where (L_{0x}, L_{0y}) are the dimensions of the object within which the phase must be reconstructed very accurately.

This qualitative discussion has contained nothing specific about the accuracy of the reconstruction corresponding to the various digitization steps. This accuracy is very sensitive to the nature of the field distribution at the objects.

If the digitization intervals along x, y, ξ, and η have been chosen, a transformation to a discrete representation of the integral in (6.10) through a DFT can be carried out as in Section 1.

In the reconstruction from Fresnel holograms, as in the reconstruction from Fourier holograms, it is necessary to take into account the masking of the object which results from the fact that the hologram is measured by detectors having a finite aperture [see (6.5)–(6.9)] .

Image Formation from Holograms. Let us examine some experiments on the digital formation of images from optical and acoustic holograms.

The optical hologram used in the experiments is a hologram recorded on photographic film at a reference-beam angle equal to the angular dimensions of the object [47]. This hologram is treated as a Fourier hologram, so that the digital reconstruction is carried out by the procedure used for reconstruction from Fourier holograms [48]. The maximum measured spatial frequency on the hologram turns out to be approximately 100 lines/mm. In order to reconcile this frequency with the digitization steps of 100 and 200 μm, the hologram in the hologram-input devices described in Section 9 is enlarged photographically by a factor of 20 (the reasons for choosing this factor and the enlargement apparatus are described in Section 9). Figure 6.1 shows a fragment of an enlarged hologram of this type. These input devices are used to enter this hologram in a computer in the form of a matrix of 512 $\times$ 512 numbers, found from measurements of the video signal on a raster of 512 $\times$ 512 elements with steps of 100 and 200 μm and with a square aperture of 25 $\times$ 25 and 100 $\times$ 100 μm, respectively. To correct for the nonlinearity of the video-signal detector, the resulting matrix of numbers is subjected to a nonlinear transformation in the first stage of the processing. In practice, this operation is similar to taking the antilogarithm since the input device measures a quantity which is approximately

Fig. 6.1

the logarithm of the transmission coefficient of the photographic film. Then a Fourier transformation of the matrix of corrected numbers is carried out. The resulting matrix of complex numbers describes the amplitude and phase distribution on the object, masked by the Fourier transform of the measuring aperture of the input device. The square moduli of these numbers, after the masking correction on the basis of (6.9), give the intensity distribution of the field at the object, i.e., an image of the object. The result of this reconstruction is shown in Fig. 6.2 for the input of a hologram with a step of 200 μm; the corresponding result for a step of 100 μm is shown in Fig. 6.3. These two images differ slightly in quality because they are found from different parts of the hologram. In the first case, the image is obtained from a part of the hologram which is 10 $\times$ 10 mm in size (on the original hologram, without enlargement). For the second image, the region is 5 $\times$ 5 mm in size. As a result, the phase noise because of the diffuseness of the objects appears to have a finer-scale structure in Fig. 6.2 than in Fig. 6.3.

As mentioned above, a masking effect arises in the reconstruction from the holograms. There are two ways to correct for this masking: by multiplying the result of the reconstruction by a correcting function [see (6.9)] or by altering the original hologram to offset the limited resolution of the detector used to record the hologram. Figure 6.4 shows the result of a reconstruction from the hologram shown in Fig. 6.1 after this alteration: a spatial filtering of the hologram in accordance with

$$\Gamma_{cor}(r, s) = \Gamma(r, s) - \frac{1}{(2N_1+1)(2N_2+1)} \sum_{k=-N_1}^{N_1} \sum_{l=-N_2}^{N_2} \Gamma(r+k, s+l).$$

(6.15)

This alteration corresponds to a multiplication of the result of an amplitude reconstruction from the hologram by the correction function

$$f(k, l) = \left[1 - \frac{\sin[\pi k(2N_1+1)/N]}{(2N_1+1)\sin(\pi k/N)}\right] \times \left[1 - \frac{\sin[\pi l(2N_2+1)/N]}{(2N_2+1)\sin(\pi l/N)}\right], \quad (6.16)$$

where N is the number of readings of the hologram, and N_1 and N_2 are parameters which determine the steepness of the correction function. In the present case, $N = 512$ and $N_1 = N_2 = 1$. It is seen from Fig. 6.4 that the alteration increases the image contrast along the edges. Furthermore, it substantially reduces the noise intensity near the center of the image; this noise is a consequence of the noise in the optical hologram itself.

The phase noise which is characteristic of images obtained from holograms can also be suppressed in part by using incoherent storage, i.e., by averaging the intensity over several images obtained from different parts of the hologram.

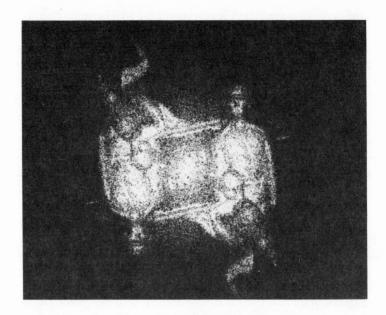

Fig. 6.2

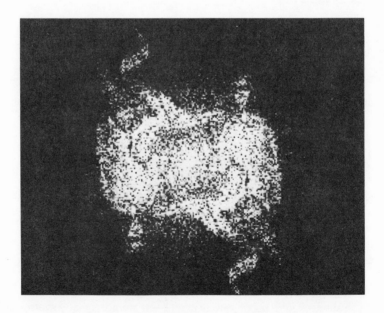

Fig. 6.3

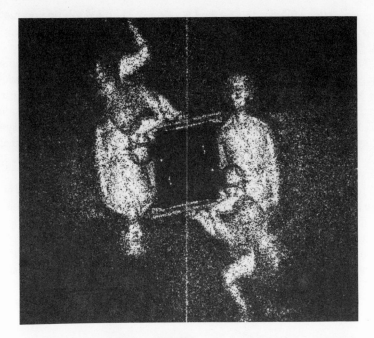

Fig. 6.4

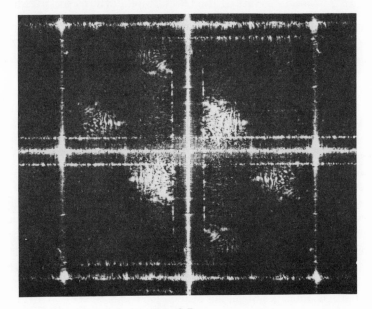

Fig. 6.5

The undesirable effect of this method is a decrease in the resolution on the image since this resolution is proportional to the area of the reconstructed hologram.

Figure 6.5 shows the result of a reconstruction from an acoustic Fourier hologram recorded by scanning a wavefront with a wavelength $\lambda = 8.3$ mm. The original objects here are three crossed cylinders. The result of the scanning is recorded on photographic film by a special device. The acoustic hologram is 12.5 × 12.5 mm in size and holds 96 × 96 independent readings. The original hologram is entered in a computer without a preliminary optical enlargement by means of an input device having a square aperture 25 μm on a side with a digitization step of 50 μm. The hologram is in the form of a square matrix of 256 × 256 numbers, which are supplemented with zeros in a symmetric way to produce a matrix of 512 × 512 numbers (see Section 7 for more details). This enlarged matrix is subjected to a Fourier transformation. The image in Fig. 6.5 has several diffraction orders because the digitization step during the entry of the hologram into the computer was several times smaller than the period of the maximum spatial frequency of the acoustic hologram. There is a substantial level of coherent noise, which can be attributed to the small dimensions of the acoustic hologram and to the low ratio of the dimensions of the object to the wavelength.

Section 7. Measurement of the Directional Patterns of Antennas

The directional pattern of an antenna is a plot of the intensity and the phase of the field radiated by an antenna in the far zone against the angle between the ray from the antenna to the measurement point and the antenna axis. This is an extremely important characteristic, and it is one of the considerations which affects the design of the antenna. After the antenna is constructed, it is usually necessary to check to see how well the actual directional pattern corresponds to the calculated pattern. For convenience and economy, it is best to measure the amplitude and the phase of the field on some surface near the antenna (in the antenna aperture) and then use a digital processor to calculate the amplitude and phase distributions in the far zone, i.e., the directional pattern. The use of digital processors to measure the directional patterns of antennas is an alternative to the analog methods [49], and it is potentially adaptable to complete automation. The relationship between the field measurements in the antenna aperture and the directional pattern of the antenna depend on the shape of the surface on which the measurements are made and the method by which the surface is scanned by the measuring receiver. In the simplest case in which the surface is a plane and the scanning is carried out in a Cartesian coordinate system, this relationship is expressed by a DFT. In more complicated cases, it may be necessary to use a discrete Fresnel transformation or some more complicated transformation. It is desirable to organize the field measurements in such a manner that the Fourier or Fresnel transformation can be used since fast algorithms are available for these cases.

Let us examine the procedure for calculating the directional pattern when a Fourier transformation is used. The dimensionality of the array of data obtained in the field measurements is usually small: It consists of matrices with something on the order of 100×100 elements. With regard to the output data, we note that in studying antennas we are usually interested in only the central lobe and its immediate vicinity, i.e., a comparatively small part of the output data. It is frequently desirable, on the other hand, to display the output data in finer detail than would be possible if the directional pattern were calculated from the field measurements alone. In order to obtain the necessary additional data points in the directional pattern, it is most convenient to use the following approach, which follows from property 15 of the DFT (Appendix I): The initial array of data is supplemented with zeros in a symmetric manner so that the ratio of the new number of data points to the old number in each direction is equal to the required number of additional data points in the directional pattern per noninterpolated reading, and the DFT transformation is carried out on this augmented array.† This procedure yields readings for the directional pattern which are interpolated in accordance with

$$
\tilde{a}\left(s_x, s_y\right) = \frac{1}{\sqrt{N_x N_y}} \sum_{r_y=0}^{N_y-1} \sum_{r_x=0}^{N_x-1} \alpha\left(r_x, r_y\right) \times
$$
$$
\times \frac{\sin\left\{\pi\left(N_x-1\right)\left[\left(s_{1x}-r_{1x}\right) L_x + s_{2x}\right]/L_x N_x\right\}}{\sin\left\{\pi\left[\left(s_{1x}-r_{1x}\right) L_x + s_{2x}\right]/L_x N_x\right\}} \times
$$
$$
\times \frac{\sin\left\{\pi\left(N_y-1\right)\left[\left(s_{1y}-r_{1y}\right) L_y + s_{2y}\right]/L_y N_y\right\}}{\sin\left\{\pi\left[\left(s_{1y}-r_{1y}\right) L_y + s_{2y}\right]/L_y N_y\right\}}, \tag{7.1}
$$

where

$$
s_x = s_{1x} L_x + s_{2x}; \quad s_y = s_{1y} L_y + s_{2y};
$$
$$
s_{1x} = 0, 1, \ldots, N_x - 1; \quad s_{1y} = 0, 1, \ldots, N_y - 1;
$$
$$
s_{2x} = 0, 1, \ldots, L_x - 1; \quad s_{2y} = 0, 1, \ldots, L_y - 1; \tag{7.2}
$$

†From the physical standpoint, the addition of zeros to the original data array corresponds to an increase in the dimensions of the antenna and to the assumption that the antenna does not radiate outside the area in which the measurements are made. By supplementing the measured data, it is possible to take into account the fact that the field outside the measurement area may actually by nonzero. This can be done, for example, by extrapolating the measurements at the boundaries and adding these extrapolated values instead of zeros to the data array.

N_x and N_y are the numbers of readings in the original data array; $\alpha(r_x, r_y)$ are the directional-pattern readings found before the original data array is supplemented with zeros; and L_x and L_y are the numbers of times the readings are duplicated.

It turns out that in a digital reconstruction of the directional patterns of antennas, the original data array to be subjected to the Fourier transformation is not filled in completely. Furthermore, as mentioned previously, very frequently it is sufficient to calculate only some of the readings of the directional pattern. In problems of this type, the truncated-FFT algorithms described in Section 2 are extremely useful.

The directional pattern which is obtained from a DFT of the original measurements may differ from the actual pattern because of a distorting effect of the detector (the receiving antenna) on the field being measured. If this effect can be assumed linear and independent of the position of the measuring antenna with respect to the antenna under study, it is possible in principle to correct the holograms for this distortion as well as for the finite aperture of the detector in the reconstruction step (Section 6).

The process of measuring the directional pattern of an antenna can be summarized as follows:

- measurement of the in-phase and squared (with respect to the reference signal) components of the field on a discrete rectangular raster in the antenna aperture and the entry of this information into the computer;
- the addition of redundant readings (for example, with zero values) to the original data array;
- a two-dimensional DFT (with the help of truncated FFT algorithms); or
- calculation of the characteristics of the directional pattern, retrieval of the data, and recording and visualization of the results.

Figures 7.1–7.3 show some results obtained by this procedure. Figure 7.1 shows the central section of the three-dimensional directional pattern of an antenna whose field was measured on a raster of 100×16 points (the antenna gain, in decibels, and the solid angle are plotted along the axes). This section is the result obtained from a one-dimensional DFT of the data array obtained by summing all 16 rows of the original matrix and adding zero values to the resultant row to achieve a length of 512 elements. In this manner, about three readings in the directional pattern can be obtained from each point in the original matrix. The one-dimensional Fourier transformation is used because only the central section is required (see Appendix II, property 7 of the two-dimensional DFT). Figure 7.2 is a three-dimensional plot of the directional pattern of the same antenna, obtained after a two-dimensional DFT of the original array of 100×16 elements, supplemented with zeros to form a $512 \times$

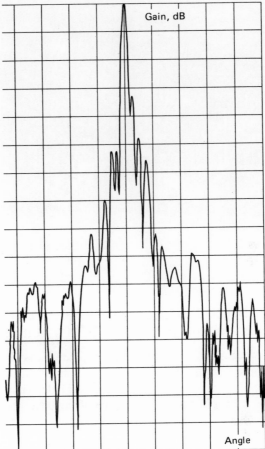

Fig. 7.1

512 array. Figure 7.3 shows another way to display the result of a calculation of a directional pattern. Here the antenna gain is shown by the blackening.

Experience with problems of this type shows that the measurements of antenna directional patterns can be carried out completely automatically with digital processors. The most promising approach is to develop a special digital complex including a minicomputer with a memory on the order of 32 kbyte, a Fourier-transformation unit (a Fourier processor), the field pickups (the measurement devices), and apparatus for recording the results. Figure 7.4 shows a possible block diagram for the complex. At the present state of computer technology, a complex of this type could furnish the directional pattern of the antenna essentially as soon as the field measurements have been made.

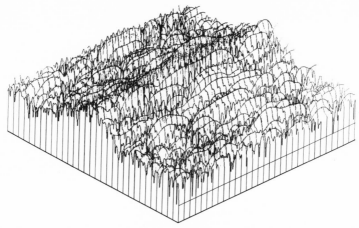

Fig. 7.2

Section 8. Digital Simulation of the Effect of Distortion during the Recording of the Holograms on the Quality of the Resulting Images

Digital Model. The property that an object diffusely scatters incident light is very important for the observation and recording of this scattered light. This property is an important factor in both physical holography and (as shown above) the synthesis of holograms. Because of this diffuse scattering, the holograms of the objects are stable with respect to distortions; any part of the

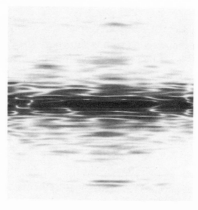

Fig. 7.3

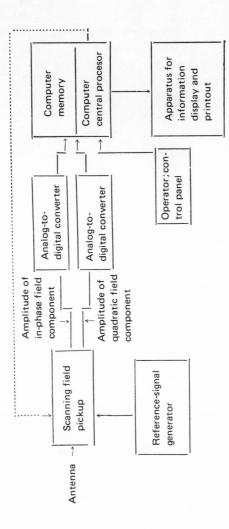

Fig. 7.4

hologram can be used to produce an image of the object; and a large dynamic range in the brightness of the object can be conveyed even though the material in which the hologram is recorded has a limited dynamic range.

This diffuseness also has its drawbacks: When diffuse objects are observed in coherent light, the macroscopic properties of the object, which are determined by the resolution of the observer, are masked by the noise associated with this diffuseness: the "speckle noise." The same noise is, of course, observed when the objects are retrieved from their holograms. The intensity of this speckle noise increases with the distortion of the field during its recording in the form of a hologram or during the reconstruction from the hologram.

We now list the basic factors which determine the distortion of the hologram as it is being recorded:

1. the finite dimensions of the hologram, i.e., the fact that it is not the entire wavefront scattered by the object but only part of it which is recorded;
2. the nonlinear distortions during the recording of the interference of the wavefront being recorded and the reference beam, including the limitation on the dynamic range of the recorded signal, the nonlinearity of the recording, and the quantization (for example, in the synthesis of holograms);
3. the digitization of the holograms (in digital synthesis); and
4. the phase distortions of the wavefront upon reconstruction from the holograms.

Other factors describe the distortions which arise during the observation of the image obtained from the holograms. The most important factors here are the finite resolution of the observer and the method by which the signal is accumulated within the resolution element (an integration of the intensity or the complex field amplitude).

To describe the effects of these factors analytically, especially their combined effects, would be a very difficult and laborious problem. Essentially the only tractable aspect of this problem is the effect of the averaging of the intensity in the plane of the image on the shape of the correlation function of the speckle noise and the signal-to-noise ratio in the image (see, for example, [50, 51]).

Digital simulation can furnish both the qualitative information required for developing approximate analytic calculation methods and specific quantitative information on the distortions and noise in the resulting image.

Figure 8.1 [65] shows a block diagram of a digital model for the recording and reconstruction of Fourier and Fresnel holograms. In this model, the object is specified by two sequences of numbers that describe the amplitude and phase of the field, respectively. These sequences are generated either by a determinate function or as a sequence of pseudorandom numbers with a Gaussian distribution and a specified energy spectrum. These sequences are produced through the

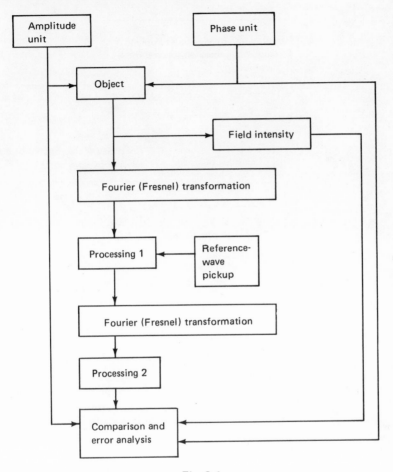

Fig. 8.1

use of the algorithm described in [44] in the following manner (Fig. 8.2): A sequence of independent pseudorandom numbers with a uniform distribution, generated by a standard device, is multiplied (element by element) by a sequence of coefficients which are the readings of the desired energy spectrum. In particular, this sequence can be supplemented with zeros, which correspond to a narrow-band spectrum of rectangular shape. The resulting sequence, masked, is subjected to the Fourier transformation, and the product is a sequence of numbers with a Gaussian distribution and the specified energy spectrum. Sequences of readings of both the amplitude and the phase can be formed in this manner. The magnitude scale of these numbers is specified as a separate parameter that determines the dispersion of these numbers. Where it is necessary to

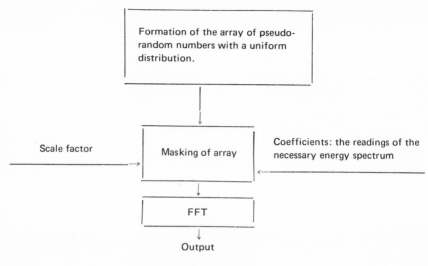

Fig. 8.2

simulate slow (macroscopic) and fast (microscopic) variations of the field phase on the object, the phase sequence can be furnished as the sum of two separate sequences. Then these two sequences are used in the "Object" unit in Fig. 8.1 to calculate sequences of values of the real and imaginary parts of the complex field amplitude at the object and a sequence of values of the field intensity, which are required for a subsequent comparison with the results of the reconstruction. These sequences can be supplemented with zeros to produce the additional points on the hologram required for a study of the effect of finite resolution during the recording of the holograms.

The resulting array of numbers is subjected to the Fourier or Fresnel transformation (depending on the problem), and the result is the array of the mathematical hologram. This array is sent to "Processing 1," where it may be transformed to simulate the hologram-recording processes. This processing unit consists of the following subunits:

1. a subunit to form the "physical hologram" by an element-by-element summation of the numbers constituting the "mathematical hologram" with the numbers corresponding to the reference wave;

2. a clipping of the signal, in which all readings of the signal which exceed some specified threshold in magnitude are replaced by this threshold (the sign is not changed);

3. quantization, in which this subunit performs a uniform quantization of the readings in terms of a specified number of levels within a specified range;

4. amplitude masking of the hologram, in which the original sequence is

multiplied by a sequence of positive numbers which determine the shape of the masking function. This subunit makes it possible to simulate apodizing and a limitation on the dimensions of the holograms;

5. phase masking, in which the original sequence is multiplied by a specified sequence of complex numbers with unit modulus. This sequence simulates the phase noise during the recording of the holograms, the flexing of the photographic film, and other effects; and

6. a smoothing subunit, in which a sliding summation of the sequence of numbers is carried out over a specified number of values. This subunit simulates the finite resolution of the medium used to record the hologram.

These subunits can be used in an arbitrary order, as specified as an instruction from the control panel of the computer. The next step is the inverse Fourier or Fresnel transformation to obtain the object. The result of this step can be subjected (in "Processing 2") to transformations to simulate the finite resolution of the device used to observe the holograms, through a sliding summation of the resulting sequence of values of the complex amplitude or intensity.

The result of the reconstruction is compared with the original object in the unit for "Comparison and error analysis." Here the dispersion and correlation function of the values of the real parts, the imaginary parts, and the square moduli of the resulting values; the dispersion and correlation function of the differences between the real parts, the imaginary parts, and the square moduli of the original and resulting sequences; and the ratio of the dispersion in the intensity of the resulting sequence to its average value (the "speckle" contrast) are found. The results of this comparison are put in the form of tables and figures.

We turn now to some results which have been obtained by this digital model.

Effect of the Finite Dimensions of the Hologram on the Reconstruction from Holograms of Objects with Diffuse and Specular Reflection. Figure 8.3 shows the variation of the reflection coefficient of the test object along the coordinates. This variation is described by a slowly varying function. Its Fourier transform (Fourier hologram) takes only five positions of the 512 positions provided in the model for conveying the hologram. If we now assign the object a random phase with independent values, we will need all 512 values for its Fourier holograms. The reconstruction from this hologram can be carried out without distortion. The limitation on the dimensions of the hologram leads to a speckle noise on the resulting object. Figure 8.4a shows the result of a reconstruction from $31/32$ of the hologram, while Fig. 8.4b shows the result obtained from $3/4$ of the hologram. We see that the noise intensifies as the dimensions of the hologram are reduced. In the case of a specular object, the reconstruction can be carried out without distortion with as little as $3/128$ of the hologram. It is not difficult to see from these figures that the noise which arises at a diffuse-reflection object is multiplicative, by which we mean that where the field is

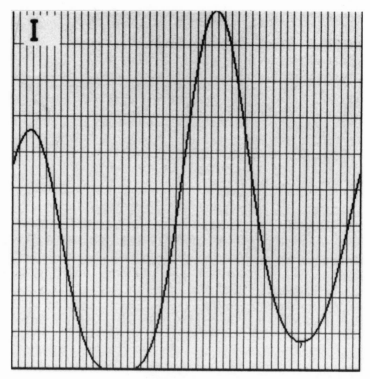

Fig. 8.3

relatively intense on the object the speckle noise is also relatively intense. This multiplicative nature of the speckle noise associated with the limited dimensions of the hologram is clearly confirmed by the fact that a plot of the standard deviation of the noise against the field intensity on the object is a straight line running parallel to the abscissa (Fig. 8.5) (line 1 corresponds to a reconstruction from $3/4$ of the hologram, line 2 to $7/8$, 3 to $15/16$, and 4 to $31/32$).

Figure 8.6 shows the standard deviation of the speckle noise as a function of the relative dimensions of the hologram for a fixed field intensity on the object in a plot of the "speckle" contrast against the relative area of the rest of the hologram.

Effect of a Limited Dynamic Range and the Quantization of the Hologram on the Reconstruction from Holograms of Diffuse and Specular Objects. The effects of a limited dynamic range and the quantization of the hologram on the reconstruction from holograms or diffuse and specular objects are completely different from the effects of the finite dimensions of the hologram. The distortion is more pronounced in the case of the specular objects. Figure 8.7 shows

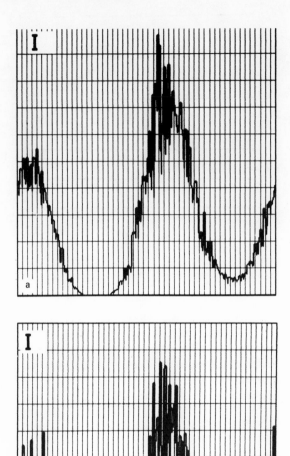

Fig. 8.4

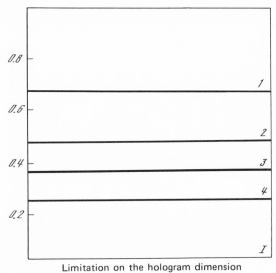

Limitation on the hologram dimension   **Fig. 8.5**

the result of a reconstruction for a diffuse object with the same distribution of the reflection coefficient as in Fig. 8.3, for the case in which the orthogonal field components of the hologram are limited to ±2.5 (a) and ±0.625 (b) of a standard deviation of the values of the components (these values have a Gaussian probability distribution). There is a noticeable increase in the speckle noise as limitation becomes more severe, but the macroscopic structure of the objects

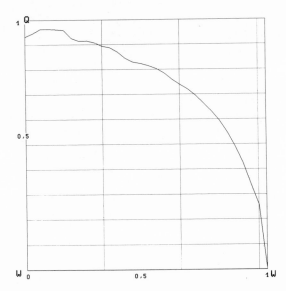

Fig. 8.6

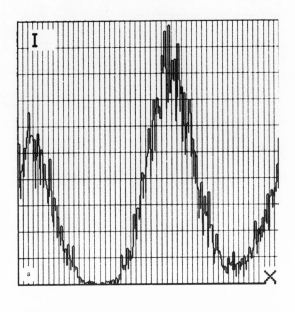

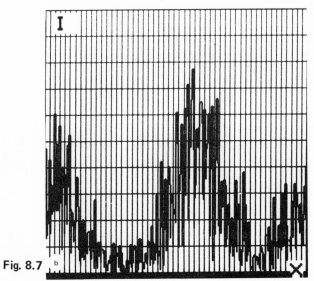

Fig. 8.7

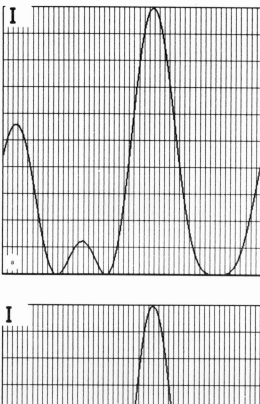

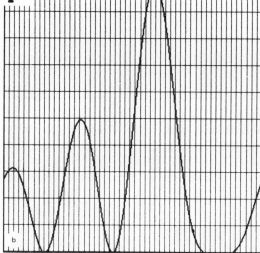

Fig. 8.8

remains. Shown for comparison in Fig. 8.8 are the results of a reconstruction from a hologram of a specular object, with the same distribution of the reflection coefficient and with the same limitation. These results show that in the case of a severe limitation of the values of the hologram for specular objects, the object is completely destroyed.

For a quantitative estimate, Fig. 8.9 shows the intensity of the speckle noise as a function of the limitation severity. Plotted along the abscissa here is the severity of the limitation on the real and imaginary parts of the hologram with respect to their standard deviation, while the quantity plotted along the ordinate is the ratio of the standard deviation of the speckle noise to the average intensity of the reconstructed field. This figure is drawn for a fixed value of the reflection coefficient.

Analogous behavior is observed in the quantization of the real and imaginary parts of the hologram: A decrease in the number of quantization levels of the hologram of a diffuse object leads to an increase in the speckle noise, but the macroscopic structure of the object is preserved [see Fig. 8.10, which shows the results of a reconstruction of the object (1) after quantization of the orthogonal field components of the hologram in the range of ±3.5 standard deviations with 128 levels (curve 2), 64 levels (3), and 32 levels (4)]. A decrease in the number of quantization levels for the hologram for a specular object leads to pronounced distortions of the macroscopic structure, as can be seen by comparing Figs. 8.11a and 8.11b, which correspond to quantization of the hologram in 128 and 16 levels, respectively. It is very instructive to follow the ratio of the standard deviation of the speckle noise to the field intensity for a diffuse object as a function of the number of quantization levels for the real and imaginary parts of the hologram (Fig. 8.12). This curve shows that as the number of quantization levels is reduced, the relative noise intensity at first increases comparatively slowly, but the increase eventually becomes extremely rapid, after about 32 levels.

Simulation of a Kinoform. The kinoform is a method for calculating and recording computer-generated holograms in which only the phase of the calculated wavefront is conveyed in the recording medium, and the amplitude is assumed constant [52-54]. The primary advantage of this method is the high diffraction efficiency of the resulting holograms: Essentially all the light incident on the hologram is used (exclusively) for the reconstruction and is not expended on the axial beam or on the conjugate image. On the other hand, since only phase information is conveyed, the hologram is seriously distorted, and this distortion could not fail to have an effect on the resulting image. If the object does not reflect diffusely, the kinoform essentially retains only the contours of the object since the equalization of the amplitudes of the different spectral components of the object causes a pronounced intensification of the higher spatial frequencies, which carry information on the contours. The diffuse object

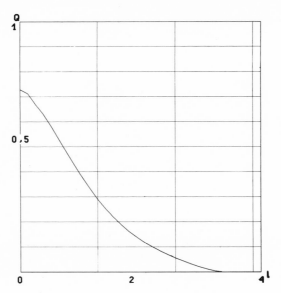

Fig. 8.9

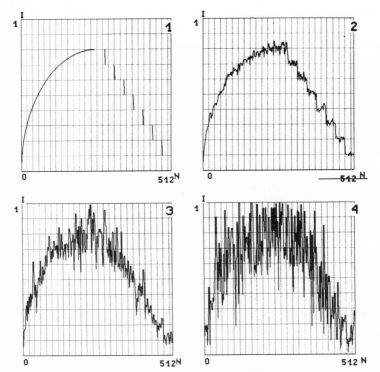

Fig. 8.10

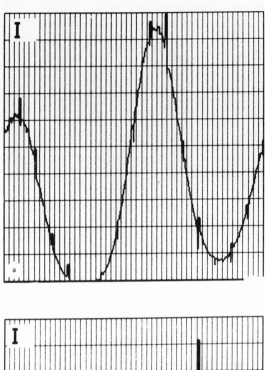

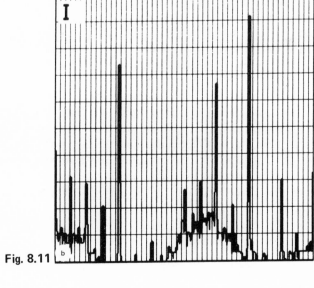

Fig. 8.11

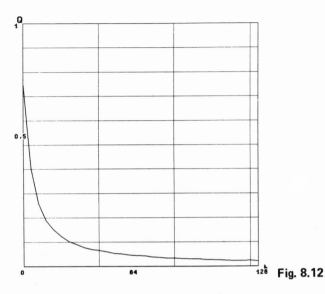

Fig. 8.12

is still conveyed, as in the case in which the hologram has a limited dynamic range, but the speckle noise is much more intense.

These comments are illustrated by Figs. 8.13 and 8.14 which show the results of a kinoform simulation.† Figure 8.13a shows the signal intensity as a step wedge. The phase of this signal is described by a sequence of pseudorandom numbers with a uniform distribution on the interval $(-\pi, \pi)$. Figure 8.13b shows the results of a reconstruction of this signal from the kinoform. Although the general behavior of the signal intensity is preserved, the step wedge has been washed out by the substantial noise level. In visual observation of the resulting images this noise would not be as apparent as in these figures, because the eye would tend to average the noise out. This circumstance can be seen by comparing the images in Fig. 8.14a (obtained in the computer from the entire hologram) and Fig. 8.14b (obtained from the same hologram, but for the case in which this hologram—like the kinoform—carries only phase information). These images were produced by using a diffuser in the synthesis of the hologram. If the diffuser is not used, the image is greatly distorted (Fig. 8.14c): Only the contours remain. Figure 8.14c shows the result of a reconstruction of a kinoform in whose synthesis there was no "phase matching" [66]. Figure 8.15 shows the result of an optical reconstruction of a physical kinoform.

---

†The Russian letters in Fig. 8.14 are the initials of the authors' institution: the Institute of Problems of Data Communications, Academy of Sciences of the USSR (translator's note).

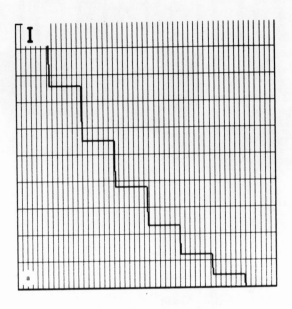

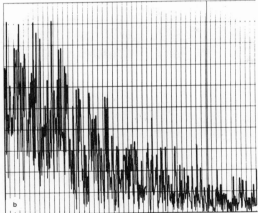

Fig. 8.13

Fig. 8.14

Fig. 8.15

3

Hardware and Software

In this chapter we will discuss the physical apparatus and the mathematical procedures used in digital holography. Technological developments in digital holography (apparatus for entering the holograms in the computer and for recording them at the computer output terminal and special digital processors to perform the transformations) and the development of the corresponding computer software are of decisive importance for the practical implementation of digital holography.

Historically, the first experiments on digital holography were carried out with whatever equipment was available. Those investigators who had access to only standard alphanumeric printers and graph plotters for computers devised "binary" holography (see, for example, [20-26]), while those who had access to more sophisticated image-display devices (this work is reviewed in [55, 56]) pursued "half-tone" digital holography [27-30, 33]. In the experiments on hologram synthesis and analysis which we have described in the present book, the apparatus used for the image input and output was half-tone apparatus with an electromechanical sweep. One of these devices, used with a Minsk 22 computer, is based on a modified facsimile apparatus. Since this device has a rather coarse raster, which does not meet the spatial–frequency requirements of the holograms, it was also necessary to photoenlarge and photoreduce the holograms. This two-step process of entering and recording the holograms will of course be made obsolete by the development of special devices for recording digital holograms, but we will describe it anyway since many of the people presently working in digital holography are still using it, and our experience may be useful. Furthermore, this method may lead to a better understanding of what these specialized devices should do. The second of the devices used, which was a step forward, allows the computer-generated holograms to be recorded immediately with rather high spatial frequencies.

The discussion of the software which follows contains a description of programs which have been written for the digital complex built around the Alpha 16 minicomputer. This computer has an operational memory of 64 kbyte (the memory access time is 1.6 μsec) and an arithmetic processor, which operates with a fixed decimal point. A special processor, which is connected to the mini-computer as an external device, is used to perform the arithmetic operations on the numbers with a floating decimal point in the complex. The average speed of the minicomputer with the external processes is about 50,000 arithmetic operations per second.

The complex also includes a 7.5-Mbyte magnetic-disk memory, a magnetic-tape memory, and the device (mentioned above) for the input and output of the images and holograms. As can be seen from these data, the computer complex used in our experiments is relatively modest in terms of calculating power. The time required to fabricate a single hologram of 1000 × 1000 elements is on the order of 0.5–1 h. The time required to solve problems involved in the hologram analysis, on the other hand, is not as long. For example, to measure the directional pattern of an antenna with this complex would take only a few minutes. The digital holography of the future will undoubtedly use specialized digital processors to perform the Fourier and other transformations required to synthesize and analyze the holograms. In this connection we believe that the software described here for an all-purpose computer will help show which types of processors with closed routines will be required.

Section 9. Hardware

Instrument Complex Using Facsimile Equipment for Hologram Input and Recording. Figure 9.1 shows a block diagram of an apparatus complex for entering the distribution of the blackening on a photographic print into a computer, for extracting the results from the computer, and for recording them on photographic film. The complex is built around a Neva facsimile transmitter and the corresponding receiver, which are used as the pickup of the video signal (the scanner) and the camera, respectively [56]. The characteristics of the complex are listed in Table 9.1.

The complex is connected to a Minsk 22 computer through a special teletype channel. The complex was designed under the assumption that the primary information carrier after the entry of the information into the computer would be magnetic tape. Since the operations memory of the Minsk 22 computer is not adequate to hold all the input and output data, a buffer memory is provided in the operational memory to assist in the exchange of information between the magnetic tape and the input–output apparatus. When the results are taken from the computer, the buffer memory is initially filled with data from the magnetic tape. Then its contents are fed to the input–output device (when the data are being fed into the computer, the operations are carried out in the opposite order). Because of this buffer, the standard length of magnetic tape is fed during the information exchange with the input–output device. A special input–output program is used to organize this part of the operation, and the standard tape

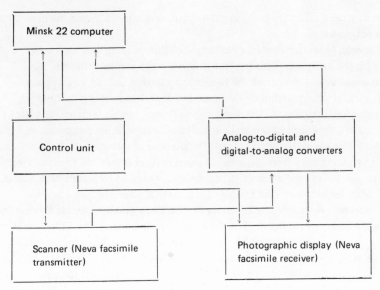

Fig. 9.1

meter of the Minsk 22 is altered so that each recording zone on the magnetic tape is 512 words long.

A special 18-digit buffer storage, with three elements of six binary digits, is used in the control apparatus to match the speed of the facsimile apparatus (2400 elements/sec at a sweep rate of 240 rpm) and the rate of the data input and output (the computer can handle no more than 800 codes/sec in the operation of the input–output program). Data on the values of the video signal are

TABLE 9.1. Basic Technical Characteristics of the Complex

Method for readout upon entry	Reflection
Input and output rate	2400 elements/sec
Raster	Rectangular, 180 × 200 μm
Aperture during readout and recording	Square, 200 × 200 μm
Measured quantity upon entry	Logarithm of the reflection coefficient of a photographic print
Recorded quantity upon output	Blackening density of a photographic negative
Number of signal-quantization levels upon input and output	64
Maximum number of image resolution elements	800 × 800

thus sent to and taken from the input–output device in a compact form at three codes per message.

A special generator controls the timing of the scanning and the exchange of data with the computer. This generator supplies the voltage powering the synchronous sweep motors of the facsimile apparatus and the timing pulses which control the operation of the analog-to-digital and the digital-to-analog converters and which go to the computer as "stop" signals. This arrangement ensures a synchronized sweep and exchange of data with the computer. A device in the control apparatus which counts the number of elements in the sweep row (it can be determined from the control panel) makes it possible to avoid mechanical methods for synchronizing the beginning of the rows by means of contacts, cams, or other breakers, and the raster is strictly regular.

Electromechanical Microdensitometer and Photographic Display System for Hologram Input and Output. The electromechanical microdensitometer and display system, manufactured by Optronics International, Chelmsford, Mass., has a rotating drum and an electromechanical scanner. The apparatus is controlled by commands from the computer. Figure 9.2 shows a block diagram of the apparatus; part a shows the microdensitometer (scanner) and part b shows the display section. The microdensitometer is an electronically controlled photometer, whose basic parts are a hollow rotating drum with a precision U-shaped carriage and an electronic control unit. On the upper arm of the carriage, which moves outside the drum, are the readout microscope objective, another aperture wheel, and a photodetector. Connected directly to the drum is an optical encoder, which generates the pulses for synchronization, readout, and the recording of data along the circumference of the drum. The motion of the carriage with the optics is independent of the drum rotation. The carriage is moved by a lead screw, which is driven in turn by a stepping motor.

The regulated light source illuminates a region of the photographic film on the drum through a square aperture. The optics system of the illuminator and the diaphragm ensure uniform illumination of this region. The optical readout head focuses the readout aperture in the illuminated part of the hologram image. The area of the illuminated region of the hologram is only 20% larger than the area of the readout aperture, so that the effects of the scattering of the light incident on the readout head are minimized. The optical magnification in the head is kept as low as possible to maximize the depth of field. Small deviations of the surface of the transparency from the drum surface thus do not affect the resolution of the apparatus. The light which passes through the film is measured by a photodetector, in this case a photomultiplier. A logarithmic amplifier takes the logarithm of the photomultiplier output current, which is proportional to the intensity of the incident light. The signal which is proportional to the light intensity transmitted by the transparency is converted into a signal which is proportional to the optical density of the transparency. The null point for the

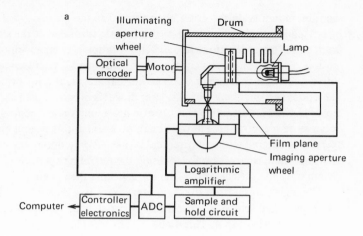

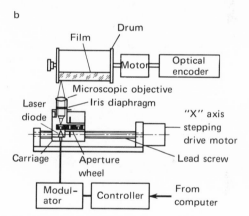

Fig. 9.2

logarithmic amplifier is provided by a reference signal on each row of the scan; this signal is obtained directly during the measurement of the total intensity of the light source. For this purpose there is a special calibration slit in the drum, running parallel to the drum axis.

The output signal from the logarithmic amplifier is sent through a digitization circuit to an analog-to-digital converter, in which the voltage proportional to the density readings on the transparency is converted into a digital signal for the computer.

The display section (Fig. 9.2b) is analogous to the microdensitometer section. The modulated light source is a laser diode, which is mounted on a carriage similar to that of the microdensitometer, but which has only a lower arm. This carriage is also moved by a lead screw. A removable light-tight cassette holding the film, mounted on the drum, is connected to the common shaft of the system by a flexible coupler. The drum of the cassette is rotated at the same time as the drum of the microdensitometer, and this rotation is independent of the carriage motion. The optical display system consists of one 5X microscope objective with a numerical aperture of 0.25, an iris diaphragm, and an aperture wheel. The optics system, along with the laser diode, is mounted on a carriage which moves outside the cassette. This cassette has a longitudinal aperture above the optics system and running in the direction in which this system moves. This aperture is covered with a shutter and is opened only during the exposure of the film. There is provision for focusing the recording aperture of the optics system onto the surface of the film for adjusting the system.

The digital signal from the computer is sent to a digital-to-analog converter and is converted into an electrical analog signal, which modulates the brightness of the laser-diode output. This diode in turn provides the appropriate illumination of the corresponding region on the film. The basic characteristics of the microdensitometer and the display section are listed in Table 9.2.

TABLE 9.2

Raster	Rectangular, with steps of 12.5, 25, 50, or 100 μm along both coordinates
Aperture	Square, with sides of 25, 50, 100, or 200 μm for readout or 12.5, 25, 50, or 100 μm for recording
Depth of field during readout and recording	From 0.05 to 0.4 mm
Density range	
Readout	0–2 OD
	0–3 OD
Recording	0–2 OD
Linearity, %	No worse than 10% (the
Blackening density – signal,	instability of the photographic
signal – blackening density	process is taken into account)
Number of quantization levels	256
Quantized quantity	
Readout	Logarithm of the transmission or reflection coefficient of a photographic transparency
Recording	Blackening density of photo-graphic film
Readout rate for the various rasters	
12.5 μm	1 rpm
25 μm	2 rpm
50, 100, and 200 μm	4 rpm
Data input rate	28,000 readings/sec
Maximum dimensions of the scanned and recorded samples	12 × 24.5 cm

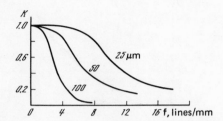

Fig. 9.3

The spatial resolution of the readout system is determined by measuring the transfer characteristic of the microdensitometer during input of data from an optical calibration slide, with a varying number of lines per millimeter, with readout apertures of 25 × 25, 50 × 50, and 100 × 100 μm. Figure 9.3 shows the transfer characteristic of the readout system according to data reported by Optronics International [57] (this is a plot of the normalized transfer ratio K against the spatial frequency f). The overall amplitude characteristic of the readout system is determined by measuring the optical density of the steps of a Kodak sensitometer wedge from 0 to 2.2 OD. The average results of these measurements for each step of the wedge are shown in Table 9.3.

The linearity of the system which converts the signal into blackening density is determined in the following manner. A sensitometer wedge with 256 steps is generated in the computer. This wedge is recorded on photographic film by the display section; the film is developed in accordance with the manufacturer's specifications; and then the film is studied with a stationary optical densitometer. The results are shown in Fig. 9.4. With this information we can correct the characteristic curve of the film for the particular physical and chemical processing.

Apparatus for Reconstruction and for Recording Images from the Computer-Generated Holograms. The reconstruction from the computer-generated holograms is carried out on the apparatus shown in Fig. 9.5 (*1* is a laser; *2*, *3*, and *4*

TABLE 9.3

Step number on sensitometer wedge	OD[a]	Measured OD in digital code	Step number on sensitometer wedge	OD[a]	Measured OD in digital code
1	0.05	9	9	1.25	153
2	0.20	25	10	1.40	168
3	0.36	43	11	1.54	186
4	0.50	62	12	1.70	203
5	0.64	79	13	1.86	218
6	0.80	100	14	2.00	234
7	0.94	118	15	2.15	252
8	1.10	138			

[a]OD = optical density.

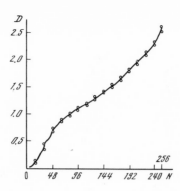

Fig. 9.4

constitute a collimator; *5* is the hologram; *6* and *7* constitute the optics system which performs the Fourier transformation; and *8* is a camera). With this apparatus, it is possible to adjust continuously the focal length of the optics system performing the Fourier transformation, and images in a given format (e.g., 24 × 36 mm) can be obtained from the computer-generated holograms with a variable digitization step $\Delta\xi$. The maximum linear dimension, X, of an image obtained from a hologram with a digitization step $\Delta\xi$, is given by

$$X = \lambda z/\Delta\xi, \tag{9.1}$$

where λ is the wavelength used in the reconstruction, and z is the focal length of the lens performing the Fourier transformation. We see from (9.1) that a change in $\Delta\xi$ is accompanied by a change in X. To keep X constant, we must simultaneously change z when $\Delta\xi$ changes. As $\Delta\xi$ is changed from 25 to 100 μm, the corresponding change in the focal length is from 100 to 400 cm. Figure 9.6 shows a simple optical system which satisfies the requirement of a continuous change in the focal length over this range and a decrease in the distance s (the distance between the outer lens L_2 and the plane in which the image is recorded, R). This system consists of two lenses; a converging lens L_1 and a diverging lens L_2. This system is known to be equivalent to a single lens L_0 with a focal length z_0 given by

$$1/z_0 = 1/z_1 + 1/z_2 - d/z_1z_2, \tag{9.2}$$

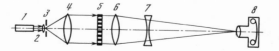

Fig. 9.5

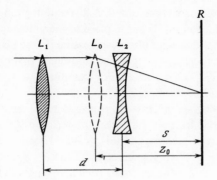

Fig. 9.6

where z_1 and z_2 are the focal lengths of L_1 and L_2, d is the distance between the lenses, and the baseline is

$$s = \frac{z_0 \, (z_1 - d)}{/z_1.} . \tag{9.3}$$

By adjusting d, the distance between the lenses, it is possible to change the focal length and baseline of the optical system. For $\Delta \xi = 50 \ \mu m$, $z_1 = 54$ cm and $z_2 = -50$ cm; for example, the distance $l = s + d$ decreases from 2 to 1.2 m for a constant image format. In the arrangement in Fig. 9.6, lenses L_1 and L_2 are mounted on a common adjustable stand, and the lens holders have a longitudinal (with respect to the stand) micrometer drive, so that lenses L_1 and L_2 can be moved without disrupting the adjustment of the system as a whole.

Apparatus for Enlarging and Reducing Holograms.†

Reduction. The apparatus used for input and output of the holograms for the Minsk 22 computer, described earlier, is capable of entering holograms with spatial frequencies no higher than 2.5 lines/mm, but the holograms typically have spatial frequencies higher by a factor of tens or hundreds. In order to enter these holograms in the computer, it is necessary to enlarge them by the appropriate factor.

The degree of enlargement required can be calculated by dividing the highest spatial frequency on the hologram by the maximum spatial frequency corresponding to the resolution of the input apparatus.

The highest spatial frequency on the hologram may be known from the arrangement for recording the hologram. It can also be determined experimentally by comparing the hologram and a diffraction grating with a known spacing on an optical bench. For this purpose, the Fourier arrangement for reconstruc-

†This apparatus was developed by D. G. Lebedev, S. K. Yefimov, and N. S. Merzlyakov.

tion from the hologram is calibrated. A diffraction grating with a known number of lines per millimeter, v_1, is put in place of the hologram in the reconstruction arrangement. In the Fourier plane we measure l_1, the distance between the diffraction peaks corresponding to the zeroth and first orders of diffraction. The parameter $a = v_1/l_1$ is determined. For example, with $v_1 = 50$ lines/mm and $l_1 = 22.5$ mm, we find $a = 2.2$ lines/mm^2.

Then the hologram under study is placed in the reconstruction device. On the image obtained from the hologram, we measure l_2, the distance from the most remote part of the image to the zeroth diffraction order. The maximum spatial frequency of the hologram is found as $v_2 = al_2$. For $l_2 = 20$ mm, for example, we have $v_2 = 44$ lines/mm. The enlargement factor k required in this case is $44/2.5 = 17.3$. Adopting $k = 20$ for simplicity, we find that the dimension of the enlarged part of the hologram in operation with matrices of numbers containing 512×512 elements is 5×5 mm, and the dimensions of the enlarged hologram are 100×100 mm. The hologram obtained from the 20X enlargement becomes the initial hologram used for the entry into the computer with the input apparatus.

For the enlargement of the holograms, a high resolution is required over the entire image field. Ordinary enlargers (the Krokus, the Belarus' 2, etc.) are not satisfactory because of the limited resolution of their objectives and also because the negative holder is not aligned precisely with respect to the axis of the objective lens. The enlargement was carried out on a standard Krokus table enlarger, but, to improve the alignment of the film with respect to the objective lens, a different objective was used and a special attachment was mounted in place of the frame holder and the bellows extension. The hologram to be enlarged is placed in this special attachment.† The negative is held between two glass plates to prevent warping by the heat from the enlarger lens.

We had three objectives at our disposal; an I5OR reproduction lens and ROZ-3M and RO-56 lenses for motion-picture cameras. According to the technical specifications, the I5OR has the highest spatial resolution, but our experience with these lenses showed that the lenses for the motion-picture cameras met our requirements better. In the reproduction of articles, journals, and other documents, the image scale is typically changed by factor of 5–10, while in the reduction and enlargement of holograms, as mentioned above, the scale is changed by a factor of 20 or more. The conditions in the recording of motion pictures are closer to our conditions.

The standard determination of the resolution of the photographic objective lenses is carried out by recording images of optical calibration slides on motion-picture film. The resolution specified in the technical specifications for the lens may thus depend on not only the objective lens itself but also the film charac-

---

†A similar attachment is shown in Fig. 9.8.

teristics. In our case, the film is Mikrat 900 film, which has a very high resolution, so that the resolution of the system as a whole is covered primarily by the objective.

To determine the resolution we used a method developed in the Central Scientific-Research Institute of Geodesy, Aerial Photography, and Cartography. This method yields the transfer ratios K for all working frequencies f (Fig. 9.7a shows the transfer ratio of the periphery of the field of view of the objective lens, while Fig. 9.7b shows the ratio at the center).

We see that the objective lenses from the motion-picture cameras have perceptible advantages over the I5OR lens. For example, for the highest frequency on the hologram, 50 lines/mm, the ROZ-3M and RO-56 lenses (curves 2 and 1) have transfer ratios of 0.85 and 0.8, respectively, while the I5OR lens (curve 3) has a transfer ratio of only 0.73 at the center of the field of view and 0.70 at the periphery. We used the ROZ-3M lens to reduce the holograms and the RO-56 lens to enlarge them. Although the latter lens is slightly poorer than the ROZ-3M, it has a focal length of 35 mm (in the tabletop arrangement used here, a 20X enlargement can be achieved only with a short-focus objective lens). The ROZ-3M and RO-56 lenses are held in holders along with the collars designed for the Konvas motion-picture camera. The image is sharpened by rotating the objective lens in the collar. The image sharpness is monitored either with the naked eye or with a magnifying glass.

Reduction. When the apparatus described above, with a digitization step $\Delta v = 200$ μm is used to record the computer-generated holograms, the angular dimensions of the image obtained in the visible-wavelength range ($\lambda = 0.63$ μm) from a hologram with a digitization step $\Delta v_x = \Delta v_y = 0.2$ mm are $2\theta_x = 2\theta_y = \lambda/\Delta v_x \simeq 6.3 \cdot 10^{-4}/0.2 = 0° \ 10'$. There are two ways to enlarge the linear and

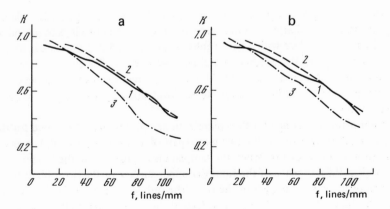

Fig. 9.7

angular dimensions of the resulting image. The first way involves the use of a coherent light source with a long wavelength, matched to the digitization step of the apparatus used in recording the holograms. The difficulties in devising an optical reconstruction arrangement (the large geometric dimensions of all the elements of the system, the need to use special optical elements and unusual recording materials, etc.) in the recording of the image and in the visualization of this image, however, force us to seek a simpler method. This simpler method is a photographic reduction of the computer-generated hologram with the use of this reduced hologram in the reconstruction.

Through calculations like those that showed us which hologram enlargement was required, in which the actual dimensions of the image are taken into account, we determine the necessary reduction of the hologram. In the present case, it is $k = 20$, and the size of the reduced hologram for the maximum number of resolution elements is 7 X 8 mm.

The apparatus for reducing the holograms is mounted on a rack held on alignment brackets mounted on a wall. The height of the rack is varied with the degree of reduction required. A microscope, with the object stage replaced by a device which holds the objective lens and the film, is mounted on the rack. The objective is pointed downward. Below the objective is a contact printer (KP8) which illuminates the reduced hologram, providing a uniform illumination of the entire hologram area. The reduced hologram is placed on the support glass plate of the KP8 printer, and another glass plate, part of the printer, is placed on top of the hologram. The printer itself is aligned in such a manner that the support glass plate is strictly horizontal. The plane of the holder for the objective lens and film in the microscope, against which the emulsion layer is pressed, is also held strictly horizontal.

For accurate alignment of the film with respect to the objective lens we use the device shown in Fig. 9.8. In the design of this device, provision was made to keep the optical axis of the objective lens perpendicular to the film plane. The construction is rigid enough to maintain sharpness when the negative is replaced. The adjustment for sharpness is made by rotating the objective lens in its collar. The film is held in place with a hard rubber plate 32 X 55 X 3 mm in size and two springs, 2. Objective 3 is held in holder 1 by screw 4. The entire holder is placed in an MBR microscope and held in place by two screws 5 and a pin 6. The film is loaded in darkness and protected from stray light during the exposure by cover 7.

A template consisting of a glass plate with an aperture was developed for the sharpness adjustment. A small metal plate (part of a safety razor blade) covers the bottom of the aperture. When the template is in position on the holder surface (Fig. 9.8), in place of the film, the upper surface of the razor blade is at the level of the emulsion layer of the film. The optical system of the microscope is initially focused so that the metal surface of the razor blade is clearly visible.

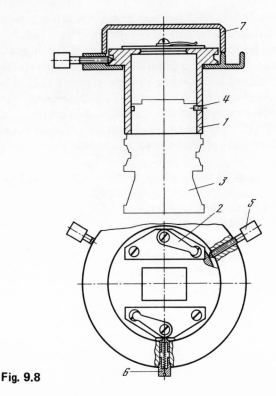

Fig. 9.8

Then the template is removed, and a sharp image of the hologram is achieved by looking through the microscope at the image of the hologram, illuminated by KP8 lamps, and rotating the collar holding the working objective. This procedure ensures that the hologram is projected onto the emulsion layer. Then the film is put in place of the template, fixed in place by the hard rubber plate, covered with the cover, and exposed.

To test the entire system we use a pattern of black and white bands alternating at a frequency of 2.5 lines/mm. This pattern was recorded on FT-20 film with a facsimile apparatus. The size of the film is 140 × 160 mm. The pattern was first reduced by a factor of 20; then it was enlarged by the corresponding factor and printed on photographic paper. This test showed that the apparatus provided sharpness over the entire image field.

Section 10. Software

The Sintez Program Library. The hologram-generation software is based on a special Sintez dialog program library, which combines programs for editing the initial data and the intermediate results, for correcting the characteristics of the input and output devices, for forming an object in accordance with a specified mathematical description, for the fast Fourier transformations, for transposing, for detecting the pseudorandom and regular phases, for multiplexing and mosaic duplication, and for recording and visualizing the calculated results. The Sintez program set is designed for an Al'pha 16 minicomputer† and can be used to generate Fourier holograms of transparencies and three-dimensional objects as well as to generate Fresnel holograms. The initial data are specified as a square matrix of 512×512 complex numbers. Figure 10.1 shows a block diagram of the program library. The operation or set of operations is chosen by the operator in a dialog mode.

Block 2 calculates the amplitude of the wavefront scattered by the object, specified as the intensity reflection (or transmission) coefficient, and it makes the original object symmetric, if necessary.

Block 3 calculates the wavefront phase at the object, which can be specified in three ways. The first way is to set the phase equal to zero. This approach corresponds to an ideally flat transparency, illuminated by a plane wave. The second way is to simulate diffuse illumination of the object. In this case the phase values are chosen from a sequence of pseudorandom numbers, having values of 0 and π with equal probabilities. In the third approach, the phase at a certain point is proportional to the distance from the given point to a plane running tangent to the object and parallel to the observation plane. As a rule, this last way for specifying the phase is the one used in generating holograms with a programmable diffuser. Block 4 multiplies the complex field amplitude at the object by a phase factor, on the basis of the Fresnel approximation. Block 5 corrects for the finite aperture of the hologram-recording apparatus (the shadow-effect correction). For this purpose the original matrix of numbers is multiplied by the following correction function before the Fourier transformation:

$$\varkappa = 1 + g \left[(i - i_0)^2 + (j - j_0)^2 \right], \tag{10.1}$$

where i and j are the indices of the row and the element, respectively (i and j take on values from 1 to 1024), $i_0 = i_{max}/2$; $j_0 = j_{max}/2$; and g is the correction factor ($g \geqslant 2$).

Block 6 uses an FFT with a joint algorithm (Section 2) to calculate the one-dimensional Fourier transformation of the array of numbers being processed. Block 7 transposes the array of numbers obtained after the one-dimensional Fourier transformation with respect to row for a subsequent transformation with respect to column. Block 8 performs a second Fourier transformation,

---

†A similar program library based on a special service program library for image processing, was developed in 1972 for the Minsk 22 computer.

1	Formation of a hologram of the pseudorandom field
2	Calculation of the field amplitude at the object
3	Calculation of the field phase at the object
4	Incorporation of the Fresnel approximation at the object
5	Shadow-effect correction
6	Fourier transformation over rows
7	Transposition
8	Fourier transformation over columns
9	Incorporation of the Fresnel approximation on the hologram
10	Multiplexing
11	Introduction of a spatial carrier
12	Encoding
13	Apodizing
14	Mosaic duplication
15	Photographic recording

Fig. 10.1

completely analogous to the transformation performed by block 6. In block 9, the resulting array is multiplied by a phase factor but takes into account the Fresnel approximation at the hologram (see Section 1). Block 10 performs the multiplexing of the calculated hologram, i.e., the spreading of some part of the hologram (usually the central part) over the entire hologram area. The same block automatically performs a repeated multiplexing, along with a simultaneous increase in the dimensions of the part of the hologram being multiplexed.

Block 11 introduces a spatial carrier into the calculated hologram in accordance with one of the following laws:

$$B_{ij} = B_{ij}(-1)^{[i/2]}, \tag{10.2}$$

$$B_{ij} = B_{ij}(-1)^{j}, \tag{10.3}$$

$$B_{ij} = B_{ij}(-1)^{j[i/2]}, \tag{10.4}$$

where $\{B_{ij}\}$ is the original array of complex numbers on the hologram, i and j have the same meaning as in (10.1), and the braces mean the largest integer. When one of these transformations is carried out, it is possible to record a complex hologram described by $\{B_{ij}\}$.

Block 12 performs a nonlinear transformation of the array of complex numbers describing the original hologram into an array of real numbers in such a manner that the magnitudes of these numbers are compatible with the characteristics of the photographic display system. The hologram is transformed in accordance with one of the following laws:

$$\tilde{B}_{ij} = kB_{ij}/(c + |B_{ij}|) + d, \tag{10.5}$$

or

$$\tilde{B}_{ij} = \log_2(|B_{ij}|m), \tag{10.6}$$

where $\{B_{ij}\}$ is the original complex array of numbers, $\{\tilde{B}_{ij}\}$ is the array of real numbers after the encoding, and k, c, m, and d are positive constants.

Block 13 superimposes an apodizing mask on the encoded hologram. Block 14 performs a mosaic duplication of the encoded hologram. The mosaics can contain from 9 to 16 identical holograms.

Block 1 is used in a simulation of the highlights on diffuse-reflecting surfaces of the objects by the programmable-diffuser method. The sequence of operations performed in block 1 is shown in Fig. 10.2. A pseudorandom-number generator initially performs a two-dimensional array of pseudorandom numbers with a uniform distribution. This array has been multiplied by the mask

$$M = (256 - |i - 256|)(256 - |j - 256|), \tag{10.7}$$

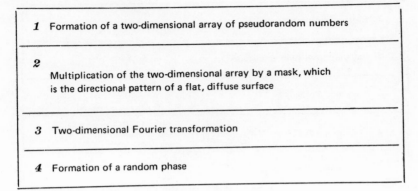

1	Formation of a two-dimensional array of pseudorandom numbers
2	Multiplication of the two-dimensional array by a mask, which is the directional pattern of a flat, diffuse surface
3	Two-dimensional Fourier transformation
4	Formation of a random phase

Fig. 10.2

where $i, j = 1, 2, \ldots, 512$. This mask is the directional pattern radiated by a flat, diffuse surface. The shape of the mask (i.e., its analytic description) determines the shape of the energy spectrum of the array of pseudorandom numbers, and it is chosen to correspond to the necessary angular distribution of the intensity of the reflected light for the diffuse surface being simulated. Then a two-dimensional discrete Fourier transformation of the masked array is carried out. Finally, a pseudorandom phase simulating the diffuseness of the surface for the illumination method used is formed from the result of the transformation. These operations constitute a universal diffuser routine, which can be used repeatedly for different objects.

This program library can be used for both floating and fixed representation of the numbers in the computer. The time required to calculate a hologram of 1024 × 1024 elements in accordance with the block diagram in Fig. 10.1 is 1 h. This time could be cut in half by using joint DFT algorithms.

The Analiz Program Library. The software for the hologram analysis is based on a special dialog service program library (the Analiz library) having the block diagram in Fig. 10.3. The Analiz library is designed to simulate the holographic process (along with the Sintez library), i.e., to study the effect of the hologram-generation methods "in their pure form" on the image quality. A second purpose of this library is to construct an image from the corresponding hologram, recorded in the visible, acoustic, and rf ranges, and to calculate the directional patterns of antennas from the measured amplitude and phase of the wavefront in the antenna aperture.

We will briefly outline the purpose of some of the blocks here. Block 2 converts the result of the entry of the hologram into the computer (block 1) into

1	Entry of hologram
2	Calculation of field amplitude
3	Editing of hologram
4	Incorporation of the Fresnel approximation at the hologram
5	Shadow-effect correction
6	Fourier row transformation
7	Transposition
8	Fourier column transformation
9	Incorporation of the Fresnel approximation of the object
10	Calculation of the modulus of the complex number
11	Encoding
12	Photographic recording

Fig. 10.3

the amplitude distribution of the hologram field. Block 3 prepares the resulting matrix of real numbers for a complex Fourier transformation. Block 4 multiplies the matrix by a phase factor which incorporates the Fresnel approximation.

The blocks for the shadow-effect correction (5), for performing the Fourier transformation (6 and 8), and for transposition (7) are completely analogous to the corresponding blocks in the Sintez library. Block 9 takes into account the Fresnel approximation at the object and is analogous to block 4. Block 10 calculates the amplitude of the reconstructed wavefront. Block 11 performs an "encoding" analogous to the operation performed by Block 12 of the Sintez programs. The result of the encoding is an image consisting of 512×512 real numbers. These are the elements of the image, which are eight-digit binary numbers or bytes, which are proportional to the image brightness at the given point and which are encoded in accordance with characteristics of the output device.

The time required to find an image from its hologram is 0.5-1 h. These calculations can be carried out with either a floating or fixed representation of the numbers in the computer.

Appendixes

I. Properties of the One-Dimensional Discrete Fourier Transformation

1. Definition. The one-dimensional discrete Fourier transformation of the sequence $\{a_k\}$, $k = 0, 1, \ldots, N - 1$, is defined by

$$\alpha_s = \frac{1}{\sqrt{N}} \sum_{k=0}^{N-1} a_k \exp\left(i2\pi \frac{ks}{N}\right) \quad s = 0, 1, \ldots, N - 1. \quad (1)$$

2. Invertibility.

$$a_k = \frac{1}{\sqrt{N}} \sum_{s=0}^{N-1} \alpha_s \exp\left(-i2\pi \frac{ks}{N}\right). \quad (2)$$

We note that

$$\frac{1}{\sqrt{N}} \sum_{s=0}^{N-1} \alpha_s \exp\left(-i2\pi \frac{ks}{N}\right) = \frac{1}{N} \sum_{s=0}^{N-1} \sum_{l=0}^{N-1} a_l \exp\left[i2\pi \frac{(l-k)s}{N}\right] =$$

$$= \sum_{l=0}^{N-1} a_l \frac{1}{N} \frac{\sin \pi (l-k)}{\sin (\pi (l-k)/N)} \exp\left[i\pi \frac{N-1}{N} (l-k)\right] = a_k, \quad (3)$$

since the function

$$\frac{1}{N} \frac{\sin \pi (l-k)}{\sin (\pi (l-k)/N)} \exp\left[i\pi \frac{N-1}{N} (l-k)\right]$$

is equal to zero for $l \neq k$ and equal to one for $l = k$.

3. Periodicity.

$$a_{r+sN} = a_{(r)\bmod N} = a_r; \quad a_{k+lN} = a_{(k)\bmod N} = a_k,$$

where $(r)_{\bmod N}$ is the remainder after r is divided by N.

This property is easy to remember if we imagine that the points r and k are measured along a circle, and the number of points in a cycle is N.

4. Conjugation. If $\{a_k\} \longleftrightarrow \{\alpha_r\}$, then

$$\{a_k^*\} \leftrightarrow \{\alpha_{N-r}^*\}, \quad \{a_{N-k}^*\} \leftrightarrow \{\alpha_r^*\}. \tag{4}$$

Then if the $\{a_k\}$ are real, i.e., if $\{a_k\} = \{a_k\}^*$, then

$$\alpha_r = \alpha_{N-r}^*. \tag{5}$$

5. Even and Odd Parity. A sequence is "even" if

$$a_k = a_{N-k}, \tag{6}$$

or it is "odd" if

$$a_k = -a_{N-k}. \tag{7}$$

With $\{a_k\} \longleftrightarrow \{\alpha_r\}$, we have

$$a_{N-k} \leftrightarrow \alpha_{N-r}, \tag{8}$$

$$a_k = -a_{N-k} \leftrightarrow \alpha_r = -\alpha_{N-r}, \tag{9}$$

$$a_k = a_{N-k}^* \leftrightarrow \alpha_r = \alpha_r^*, \tag{10}$$

$$a_k = a_{N-k} = a_k^* \leftrightarrow \alpha_r = \alpha_{N-r} = \alpha_r^*, \tag{11}$$

$$a_k = -a_{N-k} = a_k^* \leftrightarrow \alpha_r = -\alpha_{N-r} = -\alpha_r^*. \tag{12}$$

6. Parseval's Theorem. The sum of the squares of the elements of a sequence is equal to the sum of the squares to the corresponding coefficients of the DFT:

$$\sum_{r=0}^{N-1} |\alpha_r|^2 = \frac{1}{N} \sum_{r=0}^{N-1} \left[\sum_{k=0}^{N-1} a_k \exp\left(i2\pi \frac{kr}{N} \right) \right] \times$$

$$\times \left[\sum_{l=0}^{N-1} \left(a_l^* \exp\left(-i2\pi \frac{lr}{N} \right) \right) \right] =$$

$$= \frac{1}{N} \sum_{k=0}^{N-1} \sum_{l=0}^{N-1} a_k a_l^* \sum_{r=0}^{N-1} \exp\left[i2\pi \frac{(k-l)r}{N}\right] = \sum_{k=0}^{N-1} |a_k|^2. \quad (13)$$

7. Circular Cyclic Convolution. We assume

$$\{a_k\} \leftrightarrow \{\alpha_r\}, \quad \{b_k\} \leftrightarrow \{\beta_r\}. \quad (14)$$

We find the inverse DFT of $\{\alpha_r, \beta_r\}$:

$$\frac{1}{\sqrt{N}} \sum_{r=0}^{N-1} \alpha_r \beta_r \exp\left(-i2\pi \frac{kr}{N}\right) =$$

$$= \frac{1}{\sqrt{N}} \sum_{r=0}^{N-1} \left[\frac{1}{\sqrt{N}} \sum_{n=0}^{N-1} a_n \exp\left(i2\pi \frac{nr}{N}\right)\right] \times$$

$$\times \left[\frac{1}{\sqrt{N}} \sum_{m=0}^{N-1} b_m \exp\left(i2\pi \frac{mr}{N}\right)\right] \exp\left(-i2\pi \frac{kr}{N}\right) =$$

$$= \frac{1}{\sqrt{N^3}} \sum_{n=0}^{N-1} \sum_{m=0}^{N-1} \sum_{r=0}^{N-1} a_n b_m \exp\left[i2\pi \frac{(n+m-k)r}{N}\right] =$$

$$= \frac{1}{\sqrt{N}} \sum_{n=0}^{N-1} \sum_{m=0}^{N-1} a_n b_m \delta(n+m-k) = \frac{1}{\sqrt{N}} \sum_{n=0}^{N-1} a_n b_{(k-n) \bmod N}. \quad (15)$$

The product of two DFTs is thus the DFT of a circular convolution of sequences whose DFTs are in a product.

8. Circular Cyclic Displacement. We assume $\{a_k\} \longleftrightarrow \{\alpha_r\}$. We find the DFT of $a_{(k+m) \bmod N}$:

$$\frac{1}{\sqrt{N}} \sum_{k=0}^{N-1} a_{(k+m) \bmod N} \exp\left(i2\pi \frac{kr}{N}\right) =$$

$$= \frac{1}{\sqrt{N}} \sum_{l=sN+m}^{l=sN+m+N-1} a_l \exp\left[i2\pi \frac{(l-m-sN)r}{N}\right] =$$

$$= \left[\frac{1}{\sqrt{N}} \sum_{l=m}^{N+m-1} a_l \exp\left(i2\pi \frac{lr}{N}\right)\right] \exp\left(-i2\pi \frac{mr}{N}\right). \quad (16)$$

Since the summation in brackets is over the entire circle, over N values of a_l and since the frequencies of the complex exponential function are equal to l, then by changing the order of the terms in this sum we find α_r in braces. As a result we find

$$\{a_{(k+m)\bmod N}\} \leftrightarrow \left\{a_r \exp\left(-i2\pi\,\frac{mr}{N}\right)\right\}. \tag{17}$$

9. Kronecker Delta Function. We assume

$$\delta(k) = \begin{cases} 1, & k = 0, \\ 0, & k \neq 0. \end{cases} \tag{18}$$

Then

$$\delta(k-n)_{\bmod N} \leftrightarrow \frac{1}{\sqrt{N}}\exp\left[i2\pi\,\frac{(k-n)}{N}\,r\right]. \tag{19}$$

We also have

$$\delta(k_{\bmod N}) = \frac{1}{N}\sum_{l=0}^{N-1}\exp\left(i2\pi\,\frac{lk}{N}\right). \tag{20}$$

10. Spectrum of a Sinusoidal Sequence. We assume

$$a_k = \cos\left(2\pi\,\frac{r_0 k}{N} + \varphi\right) = \tfrac{1}{2}\left\{\exp\left[i\left(2\pi\,\frac{r_0 k}{N} + \varphi\right)\right]+\right.$$
$$\left.+ \exp\left[-i\left(2\pi\,\frac{r_0 k}{N} + \varphi\right)\right]\right\}, \tag{21}$$

where r_0 is the frequency of the sequence $\{a_k\}$ or the number of periods of the length of the sequence, and φ is the initial phase. We find the DFT of $\{a_k\}$:

$$\alpha_k = \frac{1}{\sqrt{N}}\sum_{k=0}^{N-1} a_k \exp\left(i2\pi\,\frac{kr}{N}\right) =$$
$$= \frac{1}{2\sqrt{N}}\left\{\sum_{k=0}^{N-1}\exp\left[i2\pi\,\frac{k(r_0+r)}{N}\right]\exp(i\varphi) + \right.$$
$$\left.+ \sum_{k=0}^{N-1}\exp\left[i2\pi\,\frac{k(r-r_0)}{N}\right]\exp(-i\varphi)\right\}. \tag{22}$$

If r_0 is an integer, then

$$a_r = \frac{\sqrt{N}}{2} \exp(i\varphi) \, \delta(r + r_0) + \frac{\sqrt{N}}{2} \exp(-i\varphi) \, \delta(r - r_0), \quad (23)$$

That is, the spectrum of a sinusoidal sequence for which the length of the sequence divided by the period is an integer consists of two Kronecker delta functions.

We now assume that r_0 is not an integer:

$$r_0 = [r_0] + \rho, \quad (24)$$

where $[r_0]$ is the greatest integer in r_0, and ρ is the remainder. Then from (22) we find

$$a_r = \tfrac{1}{2} \Bigg\{ \exp\left[i\left(\pi \frac{N-1}{N}(r+r_0) + \varphi \right) \right] \frac{(-1)^{r+[r_0]} \sin \pi\rho}{\sin[\pi(r+r_0)/N]} + \\ + \exp\left[i\left(\pi \frac{N-1}{N}(r-r_0) - \varphi \right) \right] \frac{(-1)^{r-[r_0]} \sin \pi\rho}{\sin[\pi(r-r_0)/N]} \Bigg\}. \quad (25)$$

In contrast with the continuous case, therefore, the discrete spectrum of a discrete sinusoidal signal is, in general, nonvanishing at all points.

11. Spectrum of a Discrete Square Pulse. We assume

$$\pi_{k_1}^{k_2}(k) = \begin{cases} 0, & 0 \leqslant k < k_1, \\ 1, & k_1 \leqslant k \leqslant k_2, \\ 0, & k_2 < k \leqslant N - 1. \end{cases} \quad (26)$$

We find the discrete Fourier transformation of

$$\frac{1}{\sqrt{N}} \sum_{k=0}^{N-1} \pi_{k_1}^{k_2}(k) \exp\left(i2\pi \frac{kr}{N} \right) = \frac{1}{\sqrt{N}} \sum_{k=k_1}^{k_2} \exp\left(i2\pi \frac{kr}{N} \right) = \\ = \frac{1}{\sqrt{N}} \frac{\exp[i2\pi(k_2+1)r/N] - \exp[i2\pi k_1 r/N]}{\exp[i(2\pi r/N)] - 1} = \\ = \frac{1}{\sqrt{N}} \frac{\sin[\pi r(k_2 - k_1 + 1)/N]}{\sin(\pi r/N)} \exp\left[i\pi \frac{(k_2 + k_1)r}{N} \right]. \quad (27)$$

12. Periodic Repetition. We assume

$$\{a_k\}_N \leftrightarrow \{a_r\}_N, \quad (28)$$

where N is the length of sequence $\{a_k\}$. We form the new sequence

$$\tilde{a}_l = a_{k=(l) \bmod N}, \quad l = 0, 1, \ldots, LN - 1, \tag{29}$$

where L is the number of repetitions of sequence $\{a_k\}_N$. We find the DFT of $\{\tilde{a}_l\}$:

$$\text{DFT} \{\tilde{a}_l\}_{LN} = \frac{1}{\sqrt{LN}} \sum_{l=0}^{LN-1} \tilde{a}_l \exp\left(i2\pi \frac{lr}{LN}\right). \tag{30}$$

We adopt the notation

$$l = l_1 N + l_2, \text{where } l_1 = 0, 1, \ldots, L - 1; l_2 = 0, 1, \ldots, N - 1,$$

We break up the sum into two parts, over l_1 and l_2:

$$\text{DFT} \{\tilde{a}_l\}_{LN} = \frac{1}{\sqrt{LN}} \sum_{l_1=0}^{L-1} \sum_{l_2=0}^{N-1} \tilde{a}_{l_1 N + l_2} \exp\left[i2\pi \frac{(l_1 N + l_2) r}{LN}\right] =$$

$$= \frac{1}{\sqrt{LN}} \sum_{l_1=0}^{L-1} \sum_{l_2=0}^{N-1} a_{l_2} \exp\left(i2\pi \frac{l_1 r}{L}\right) \exp\left(i2\pi \frac{l_2 r}{LN}\right) =$$

$$= \frac{1}{\sqrt{LN}} \sum_{l_1=0}^{L-1} \exp\left(i2\pi \frac{l_1 r}{L}\right) \sum_{l_2=0}^{N-1} a_{l_2} \exp\left(i2\pi \frac{l_2 r}{LN}\right) =$$

$$= \sqrt{\frac{L}{N}} \sum_{l_2=0}^{N-1} a_{l_2} \exp\left(i2\pi \frac{l_2 r}{LN}\right) \delta\left(r_{\bmod L}\right). \tag{31}$$

We adopt the notation

$$r = r_1 L + r_2; r_1 = 0, 1, \ldots, N - 1; r_2 = 0, 1, \ldots, L - 1,$$

We find

$$\text{DFT} \{\tilde{a}_l\}_{LN} = \tilde{a}(r) = \tilde{a}(r_1 L + r_2) = \sqrt{L} a(r_1) \delta(r_2), \tag{32}$$

That is, in the case of a periodic continuation of a sequence, the discrete spectrum can be found from an integral number of periods by increasing the spacing of the elements in the original spectrum equal to the number of repetition periods.

13. Increasing the Spacing between Elements by a Number of Positions.
From Theorem 12 we find the inverse theorem: An increase in the spacing of

the elements leads to a periodic repetition of their spectrum. If then

$$\{\tilde{a}_l\}_{LN} = \{a_{l_1}\delta(l_2)\} \leftrightarrow \sqrt{\frac{1}{L}}\{\alpha(r_2)\}_{LN}, \tag{33}$$

where

$$l = l_1 L + l_2; \quad l_1 = 0,1, \ldots, N-1; \quad l_2 = 0,1, \ldots, L-1;$$
$$r = r_1 N + r_2; \quad r_1 = 0,1, \ldots, L-1; \quad r_2 = 0,1, \ldots, N-1.$$

14. Addition of Zeros.

From the sequence

$$\{a_k\}_N \leftrightarrow \{\alpha_r\}_N \tag{34}$$

we find the new sequence

$$\{\tilde{a}_l\}_{LN} = \{\tilde{a}_{l_1N+l_2}\}_{LN} = \{a_{l_2}\delta(l_1)\}_{LN},$$
$$l_1 = 0, 1, \ldots, \quad L-1; \quad l_2 = 0, 1, \ldots, N-1; \tag{35}$$

in other words, we add $(L-1)N$ zeros on the right of the original sequence. We find DFT $\{\tilde{a}_l\}_{LN}$, using Theorem 7 regarding convolution. We can obviously write

$$\{\tilde{a}_l\}_{LN} = \{a_{l_2}\pi_0^{N-1}(l)\}_{LN} \leftrightarrow \frac{1}{\sqrt{LN}} \text{DFT}\{a_{l_2}\}_{LN} \circledast \text{DFT}\{\pi_0^{N-1}(l)\} =$$

$$= \sqrt{L}\alpha(r_1)\,\delta(r_2) \circledast \frac{1}{\sqrt{LN}}\frac{\sin(\pi r/L)}{\sin(\pi r/LN)}\exp\left(i\pi\,\frac{N-1}{LN}\right) =$$

$$= \frac{1}{LN}\sqrt{L}\sum_{r=0}^{LN-1}\alpha(r_1)\,\delta(r_2)\frac{\sin(\pi/L)(s-r)_{\text{mod }LN}}{\sin(\pi/LN)(s-r)_{\text{mod }LN}}\times$$

$$\times \exp\left[i\pi\,\frac{N-1}{LN}(s-r)_{\text{mod }LN}\right] =$$

$$= \sqrt{\frac{1}{L}}\sum_{r_1=0}^{N-1}\alpha(r_1)_N\frac{\sin(\pi/L)(s_1L+s_2-r_1L)_{\text{mod }LN}}{\sin(\pi/LN)(s_1L+s_2-r_1L)_{\text{mod }LN}}\times$$

$$\times \exp\left[i\pi\,\frac{N-1}{LN}(s_1L+s_2-r_1L)_{\text{mod }LN}\right], \tag{36}$$

In other words, the spectrum of a sequence to which zeros have been added is equal to the spectrum of the original sequence after the spacing between elements has been increased, interpolated in the gaps by the function

$$\frac{\sin (\pi/L) \, [(s_1 - r_1) L + s_2]_{\text{mod } LN}}{\sin (\pi/LN) \, [(s_1 - r_1) L + s_2]_{\text{mod } LN}} \times$$

$$\times \exp \left\{ i\pi \frac{N-1}{LN} \, [(s_1 - r_1) L + s_2]_{\text{mod } LN} \right\}, \qquad (37)$$

Then

$$s = s_1 L + s_2; \; s_1 = 0,1, \ldots, N - 1; \; s_2 = 0,1, \ldots, L - 1.$$

15. Symmetric Addition of Zeros. From

$$\{ \hat{a}_k \}_N \leftrightarrow \{ a_r \}_N \qquad (38)$$

where N is an even integer, we form the sequence

$$\{ \bar{a}_l \}_{LN} = \{ a_{l_2} (1 - \pi_{N/2}^{LN-N/2} (l)) \}_{LN}$$

and find its DFT. We first determine the DFT of $(1 - \pi_{N/2}^{LN-N/2}(l))_{LN}$:

$$\text{DFT} (1 - \pi_{N/2}^{LN-N/2} (l))_{LN} = \frac{1}{\sqrt{LN}} \left[\sum_{l=0}^{N/2-1} \exp \left(i2\pi \frac{lr}{LN} \right) + \right.$$

$$\left. + \sum_{l=LN-N/2+1}^{LN-1} \exp \left(i2\pi \frac{lr}{LN} \right) \right] =$$

$$= \frac{1}{\sqrt{LN}} \frac{\exp (i\pi r/L) - 1 + 1 - \exp [- i\pi (N - 2) r/LN]}{\exp (i2\pi r/LN) - 1} =$$

$$= \frac{1}{\sqrt{LN}} \frac{\sin [\pi (N - 1) r/LN]}{\sin [\pi r/LN]} . \qquad (39)$$

Then the spectrum of the sequence $\{ \tilde{a}_l \}$ is

$$\tilde{a} (s) = \frac{1}{\sqrt{L}} \sum_{r_1=0}^{N-1} a(r_1) \frac{\sin \{ \pi [(N - 1)/LN] [(s_1 - r_1) L + s_2] \}}{\sin \{ \pi [(s_1 - r_1) L + s_2]/LN \}} . \qquad (40)$$

Theorems 14 and 15 may be called the "sampling theorems" or "reading theorems" for the discrete Fourier transformation.

16. Theorem on Permutation. This is an analog of the theorem on scales for the integral Fourier transformation. We assume

$$\{a_k\}_N \leftrightarrow \{\alpha_r\}_N. \tag{41}$$

We construct the sequence

$$\{b_l\}_N = \{a_{(pk) \bmod N}\}_N, \tag{42}$$

where p is an integer which is not a divisor of N. For example, for $N = 9$ and $p = 5$, the correspondence between b_l and a_k is as follows:

b_l	b_0	b_1	b_2	b_3	b_4	b_5	b_6	b_7	b_8
a_k	a_0	a_5	a_1	a_6	a_2	a_7	a_3	a_8	a_4.

We find DFT of $\{b_k\}_N$:

$$\beta_r = \frac{1}{\sqrt{N}} \sum_{k=0}^{N-1} b_k \exp\left(i2\pi \frac{kr}{N}\right) =$$

$$= \frac{1}{\sqrt{N}} \sum_{k=0}^{N-1} a_{(pk) \bmod N} \exp\left(i2\pi \frac{kr}{N}\right). \tag{43}$$

We assume that k_1 is such that $(pk_1)_{\bmod N} = 1$. (In the example here, $k_1 = 2$.) We use the substitution of variables

$$k = (k_1 l)_{\bmod N}; \quad l = 0, 1, , \ldots, N - 1. \tag{44}$$

This change is required so that we can distinguish, in the subscript on a, a variable which changes sequentially† from 0 to $N - 1$:

$$\beta_r = \frac{1}{\sqrt{N}} \sum_{l=0}^{N-1} a_{(p(k_1,l) \bmod N) \bmod N} \exp\left[i2\pi \frac{r(k_1 l)_{\bmod N}}{N}\right] =$$

$$= \frac{1}{\sqrt{N}} \sum_{l=0}^{N-1} a_{p(k_1,l) \bmod N} \exp\left[i2\pi \frac{l(rk_1)_{\bmod N}}{N}\right] =$$

$$= \frac{1}{\sqrt{N}} \sum_{l=0}^{N-1} a_l \exp\left[i2\pi \frac{l(rk_1)_{\bmod N}}{N}\right] = \alpha\left[(rk_1)_{\bmod N}\right], \tag{45}$$

†It is not difficult to prove the validity of these results of operations with numbers which are taken in modulus. As an example we consider the operation of putting numbers within the modulus sign. We assume $s = s_1 N + s_2$, that is, $(s)_{\bmod N} = s_2$. Obviously, $(ls)_{\bmod N} = (ls_1 N + ls_2)_{\bmod N} = (ls_2)_{\bmod N} = (l(s)_{\bmod N})_{\bmod N}$.

In other words, the permutation of the original sequence leads to some permutation of its DFT. For $N = 9$ and $p = 5$, this permutation is

$$
\begin{array}{ccccccccc}
\beta_r & \beta_0 & \beta_1 & \beta_2 & \beta_3 & \beta_4 & \beta_5 & \beta_6 & \beta_7 & \beta_8 \\
\alpha_{rk_1} & \alpha_0 & \alpha_2 & \alpha_4 & \alpha_6 & \alpha_8 & \alpha_1 & \alpha_3 & \alpha_5 & \alpha_7.
\end{array}
$$

The numbers p and k_1 are inverse with respect to the modulus of N: $(pk_1)_{\mathrm{mod}\ N} = 1$ (see the analogous property for the integral Fourier transformation).

17. Mean Value Theorem.

$$
\frac{1}{N} \sum_{k=0}^{N-1} a_k = \frac{1}{\sqrt{N}} \alpha_0. \tag{46}
$$

The zeroth coefficient of the DFT of ths sequence $\{a_k\}$ is proportional to the mean value over the sequence.

18. Theorem on the Mean Values of the Even and Odd Elements of the Sequence. For even N,

$$
\alpha_{N/2} = \frac{1}{\sqrt{N}} \sum_{k=0}^{N-1} a_k \exp(i\pi k) = \frac{1}{\sqrt{N}} \sum_{k=0}^{N-1} (-1)^k a_k =
$$

$$
= \frac{1}{\sqrt{N}} \left(\sum_{k=0}^{N/2-1} (a_{2k} - a_{2k+1}) \right). \tag{47}
$$

Hence

$$
\frac{\alpha_0 + \alpha_{N/2}}{\sqrt{N}} = \frac{2}{N} \sum_{k=0}^{N/2-1} a_{2k}, \tag{48}
$$

$$
\frac{\alpha_0 - \alpha_{N/2}}{\sqrt{N}} = \frac{2}{N} \sum_{k=0}^{N/2-1} a_{2k+1}. \tag{49}
$$

II. Properties of the Two-Dimensional Discrete Fourier Transformation

1. Definition. Just as the one-dimensional sampling theorem generates a one-dimensional DFT, the two-dimensional sampling theorem generates a two-dimensional DFT. We will consider only that two-dimensional DFT which follows from the two-dimensional integral Fourier transformation in Cartesian coordinates:

$$a_{r,s} = \frac{1}{\sqrt{NM}} \sum_{k=0}^{N-1} \sum_{l=0}^{M-1} a_{k,l} \exp\left[i2\pi\left(\frac{kr}{N} + \frac{ls}{M}\right)\right]. \tag{1}$$

This transformation is convenient because it is factored into two one-dimensional DFTs as a consequence of the way in which exponential functions multiply.

The inverse two-dimensional DFT is written

$$a_{k,l} = \frac{1}{\sqrt{NM}} \sum_{k=0}^{N-1} \sum_{s=0}^{M-1} a_{r,s} \exp\left[-i2\pi\left(\frac{kr}{N} + \frac{ls}{M}\right)\right]. \tag{2}$$

2. Periodicity. If, for given $\{a_k\}$, the result of their one-dimensional DFT can be thought of as specified at points on a circle, then in the case of the two-dimensional DFT we can speak in terms of points specified on a torus:

$$a_{k,l} = a_{(k)\bmod N,\,(l)\bmod M}; \quad a_{r,s} = a_{(r)\bmod N,\,(s)\bmod M}. \tag{3}$$

3. Conjugation. From $\{a_{k,l}\}_{NM} \longleftrightarrow \{\alpha_{r,s}\}_{NM}$ we find

$$\{a_{k,l}^{*}\} \leftrightarrow \{\alpha_{N-r,\,M-s}^{*}\}. \tag{4}$$

If $\{a_{k,l}\} = \{a_{k,l}^{*}\}$, i.e., $\{a_{k,l}\}_{NM}$ are a purely real array of numbers, then

$$\{\alpha_{r,s}\}_{NM} = \{\alpha_{N-r,\,M-s}^{*}\}. \tag{5}$$

The inverse relations also hold:

$$\{\alpha_{r,s}^{*}\} \leftrightarrow \{a_{N-k,\,M-l}^{*}\}, \quad \{\alpha_{r,s}\} = \{\alpha_{r,s}^{*}\} \leftrightarrow \{a_{k,l}\} = \{a_{N-k,\,M-l}^{*}\}. \tag{6}$$

4. Even and Odd Parity. From the definition of the two-dimensional DFT we have

$$\{a_{-k,-l}\} = \{a_{N-k,\,M-l}\} \leftrightarrow \{a_{-r,-s}\} = \{\alpha_{N-r,\,M-s}\}, \tag{7}$$

$$\{a_{-k,\,l}\} = \{a_{N-k,\,l}\} \leftrightarrow \{\alpha_{-r,\,s}\} = \{\alpha_{N-r,\,s}\}. \tag{8}$$

From (4)–(8) we find a relation for sequences which are even in two dimensions:

$$\{a_{k,l}\} = \{a_{N-k,\,M-l}\} \leftrightarrow \{\alpha_{r,s}\} = \{\alpha_{N-r,\,M-s}\} \tag{9}$$

For sequences which are odd in two dimensions,

$$\{a_{k,l}\} = \{-a_{N-k,\,M-l}\} \leftrightarrow \{\alpha_{r,s}\} = \{-\alpha_{N-r,\,M-s}\}. \tag{10}$$

Analogously, for sequences which are, respectively, even and odd in one dimension,

$$\{a_{k,l}\} = \{a_{N-k,l}\} \leftrightarrow \{\alpha_{r,s}\} = \{\alpha_{N-r,s}\}, \tag{11}$$

$$\{a_{k,l}\} = \{-a_{N-k,l}\} \leftrightarrow \{\alpha_{r,s}\} = \{-\alpha_{N-r,s}\}. \tag{12}$$

Finally,

$$\{a_{k,l}\} = \{a_{N-k,\,M-l}\} = \{a_{k,l}^{*}\} \leftrightarrow \{\alpha_{r,s}\} = \{\alpha_{N-r,\,M-s}\} =$$
$$= \{\alpha_{N-r,\,M-s}^{*}\}, \tag{13}$$

$$\{a_{k,l}\} = \{-a_{N-k,\,M-l}\} = \{a_{k,l}^{*}\} \leftrightarrow \{\alpha_{r,s}\} =$$
$$= \{-\alpha_{N-r,\,M-s}\} = \{\alpha_{N-r,\,M-s}^{*}\}, \tag{14}$$

In other words, the DFT of a real sequence which is even in two dimensions is a real sequence which is even in two dimensions, and the DFT of a real sequence which is odd in two dimensions is a purely imaginary sequence which is odd in two dimensions.

5. Parseval's Theorem. As for the one-dimensional DFT, the sum of the squares of the Fourier coefficients of the two-dimensional signals is equal to the sum of the squares of the readings of the signal:

$$\sum_{r=0}^{N-1} \sum_{s=0}^{M-1} |\alpha_{rs}|^2 = \sum_{k=0}^{N-1} \sum_{l=0}^{M-1} |a_{kl}|^2. \tag{15}$$

The derivation of this result is analogous to that in Appendix I for the one-dimensional case.

6. Two-Dimensional Cyclic Convolution. We assume

$$\{a_{k,l}\} \leftrightarrow \{\alpha_{r,s}\}, \quad \{b_{k,l}\} \leftrightarrow \{\beta_{r,s}\}. \tag{16}$$

Then

$$\frac{1}{\sqrt{NM}} \sum_{r=0}^{N-1} \sum_{s=0}^{M-1} \alpha_{r,s}\beta_{r,s} \exp\left[-i2\pi\left(\frac{kr}{N} + \frac{ls}{M}\right)\right] =$$
$$= \frac{1}{\sqrt{NM}} \sum_{n=0}^{N-1} \sum_{m=0}^{M-1} a_{n,m} b_{(k-n)\bmod N,\,(l-m)\bmod M}. \tag{17}$$

Since the two-dimensional DFT is factored into two one-dimensional DFTs, the derivation of this equation is precisely the same as the derivation of the corresponding one-dimensional equation in Appendix I.

7. Cyclic Displacement. As a consequence of the periodicity of a two-dimensional DFT, the multiplication of the Fourier components by a complex exponential function with a linearly increasing argument in the two-dimensional case, as in the one-dimensional case, leads to a cyclic displacement of the corresponding signal:

$$\left\{a_{r,\,s}\exp\left[-i2\pi\left(\frac{nr}{N}+\frac{ms}{M}\right)\right]\right\}\leftrightarrow\left\{a_{(k+n)\bmod N,\ (l+m)\bmod M}\right\}. \quad (18)$$

8. Periodic Duplication and Addition of Zeros. We assume

$$\{a_{k,l}\}_{N,M}\leftrightarrow\{a_{r,\,s}\}_{N,M} \quad (19)$$

and

$$n = n_1 N + n_2;\ m = m_1 M + m_2;\ n_1 = 0,\,1,\,\ldots,\,L_n-1;$$
$$n_2 = 0,\,1,\,\ldots,\,N-1;\quad m_1 = 0,\,1,\,\ldots,\,L_{m-1};$$
$$m_2 = 0,\,1,\,\ldots,\,M-1. \quad (20)$$

We find the DFT of the sequence $\{a_{n_2,m_2}\}_{L_n N, L_m M}$:

$$\tilde{a}_{p,\,q} = \frac{1}{\sqrt{L_n L_m NM}}\sum_{n_1=0}^{L_n-1}\sum_{n_2=0}^{N-1}\sum_{m_1=0}^{L_m-1}\sum_{m_2=0}^{M-1}a_{n_2,\,m_2}\times$$

$$\times\exp\left\{i2\pi\left[\frac{p(n_1 N + n_2)}{L_n N}+\frac{q(m_1 M + m_2)}{L_m M}\right]\right\}=$$

$$=\frac{1}{\sqrt{L_n L_m NM}}\sum_{n_1=0}^{L_n-1}\sum_{n_2=0}^{N-1}\sum_{m_1=0}^{L_m-1}\sum_{m_2=0}^{M-1}a_{n_2,\,m_2}\exp\left(i2\pi\,\frac{pn_1}{L_n}\right)\times$$

$$\times\exp\left(i2\pi\frac{pn_2}{L_n N}\right)\exp\left(i2\pi\,\frac{qm_1}{L_m}\right)\exp\left(i2\pi\frac{qm_2}{L_m M}\right). \quad (21)$$

We adopt the notation

$$p = p_1 L_n + p_2;\qquad q = q_1 L_m + q_2;$$
$$p_1 = 0,\,1,\,\ldots,\,N-1;\qquad p_2 = 0,\,1,\,\ldots,\,L_n-1;$$
$$q_1 = 0,\,1,\,\ldots,\,M-1;\qquad q_2 = 0,\,1,\,\ldots,\,L_m-1, \quad (22)$$

and we consider the sum over n_1 and m_1 in (21):

$$\sum_{n_1=0}^{L_n-1} \sum_{m_1=0}^{L_m-1} \exp\left(i2\pi \frac{pn_1}{L_n}\right) \exp\left(i2\pi \frac{qm_1}{L_m}\right) = L_m L_n \delta(p_2)\, \delta(q_2) =$$

$$= L_n L_m \delta((p)_{\text{mod } L_n})\, \delta((q)_{\text{mod } L_m}). \tag{23}$$

As a result we find

$$\tilde{a}_{p,q} = \sqrt{\frac{L_n L_m}{NM}} \sum_{n_2=0}^{N-1} \sum_{m_2=0}^{M-1} a_{n_2 m_2} \exp\left(i2\pi \frac{p_1 n_2}{N}\right) \times$$

$$\times \exp\left(i2\pi \frac{q_1 m_2}{M}\right) \delta(p_2)\, \delta(q_2) = \sqrt{L_n L_m}\, a_{p_1 q_1} \delta(p_2)\, \delta(q_2). \tag{24}$$

As in the one-dimensional relation, the periodic duplication of the original sequence leads to the corresponding increase in the spacing of the discrete components of its DFT.

It is not difficult to show that the addition of zeros to the sequence generates a DFT consisting of the interpolated components of the spectrum of the original sequence, with the spacing of these components increased.

7. Mean Value Theorem.

$$a_{0,0} = \frac{1}{\sqrt{NM}} \sum_{k=0}^{N-1} \sum_{l=0}^{M-1} a_{k,l}, \tag{25}$$

$$a_{0,s} = \frac{1}{\sqrt{NM}} \sum_{l=0}^{M-1} \left(\sum_{k=0}^{N-1} a_{k,l}\right) \exp\left(i2\pi \frac{ls}{M}\right), \tag{26}$$

$$a_{0,M/2} = \frac{1}{\sqrt{NM}} \sum_{l=0}^{M-1} (-1)^l \sum_{k=0}^{N-1} a_{kl}, \tag{27}$$

$$a_{N/2,0} = \frac{1}{\sqrt{NM}} \sum_{k=0}^{N-1} (-1)^k \sum_{l=0}^{M-1} a_{kl}, \tag{28}$$

$$a_{N/2,M/2} = \frac{1}{\sqrt{NM}} \sum_{k=0}^{N-1} \sum_{l=0}^{M-1} (-1)^{k+l} a_{k,l}. \tag{29}$$

References

1. D. Gabor, A new microscopic principle, *Nature,* **161,** 777-778 (1948).
2. D. Gabor, Microscopy by reconstructed wavefronts, 1, *Proc. Roy. Soc.,* **A197,** 454-487 (1949).
3. W. L. Bragg, Microscopy by wavefront reconstruction, *Nature,* **166,** 399-403 (1950).
4. G. W. Stroke, *An Introduction to Coherent Optics and Holograpy,* Academic Press, New York (1966).
5. S. A. Benton, Holographic displays—a review, *Opt. Engin.,* **14,** No. 5, 402-407 (1975).
6. D. Gabor, Light and information, in: *Progress in Optics,* 1, ed. E. Wolf, Amsterdam (1961), pp. 109-153.
7. E. N. Leith and J. Upatnieks, New techniques in wavefront reconstruction, *J. Opt. Soc. Am.,* **51,** 1469-1473 (1961).
8. Yu. N. Denisyuk, Photographic reconstruction of the optical properties of an object in its own scattered radiation field, *Dokl. Akad. Nauk. SSSR,* **144,** 1275-1279 (1962).
9. E. N. Leith and J. Upatnieks, Reconstructed wavefronts and communication theory, *J. Opt. Soc. Am.,* **52,** 1123-1130 (1962).
10. Yu. N. Denisyuk, Manifestation of the Optical Properties of an Object in the Wavefront of Light Scattered by This Object. Author's abstract, Candidate's dissertation [in Russian], Leningrad (1963).
11. E. N. Leith and J. Upatnieks, Wavefront reconstruction with continuous-tone objects, *J. Opt. Soc. Am.,* **53,** 1377-1381 (1963).
12. T. S. Huang, Digital holography, *Proc. IEEE,* **59,** No. 9, 1335-1346 (1971).
13. D. Gabor, Holography of the "Whole picture," *New Scientist,* **29,** No. 4, 74-78 (1966).
14. Yu. N. Denisyuk, Photography which reproduces a complete illusion of reality, *ZhNiPFiK,* **11,** 46-56 (1966).
15. B. Gold and C. M. Rader, *Digital Processing of Signals,* McGraw-Hill, New York (1969).
16. I. J. Good, The interaction algorithm and practical Fourier analysis, *J. Roy. Statist. Soc., Ser. B,* **20,** 361-372 (1958).
17. D. C. Chu, J. R. Fienup, and J. W. Goodman, Multiemulsion on-axis computer generated hologram, *Appl. Opt.,* **12,** No. 7, 1386-1388 (1973).
18. D. C. Chu and J. R. Fienup, Recent approaches to computer-generated color holograms, *Optical Engineering,* **13,** No. 3, 189-195 (1974).
19. J. R. Fienup and J. W. Goodman, New ways to make computer-generated color holograms, *Nouv. Rev. Optique,* **5,** 269-275 (1974).
20. B. R. Brown and A. W. Lohmann, Complex spatial filtering with binary masks, *Appl. Opt.,* **5,** No. 6, 967-969 (1966).
21. A. W. Lohmann and D. P. Paris, Binary Fraunhofer holograms, generated by computer, *Appl. Opt.,* **6,** No. 10, 1739-1748 (1967).

22. A. W. Lohmann, D. P. Paris, and H. W. Werlich, A computer generated spatial filter applied to code translation, *Appl. Opt.* **6**, No. 6, 1139–1140 (1967).
23. B. R. Brown, A. W. Lohmann, and D. P. Paris, Computer-generated optical-matched filtering, *Opt. Acta,* **3**, 377–379 (1966).
24. A. W. Lohmann and D. P. Paris, Computer-generated spatial filter for coherent optical data processing, *Appl. Opt.,* **7**, No. 4, 651–655 (1968).
25. B. R. Brown and A. W. Lohmann, Computer-generated binary holograms, *IBM J. Res. Dev.,* **13**, No. 2, 160–168 (1969).
26. A. W. Lohmann, Matched filtering with self-luminous objects, *Appl. Opt.* **7**, No. 3, 561–563 (1968).
27. W. H. Lee, Computer-Generation of Holograms and Spatial Filters, Sc. D. dissertation, Department of Electrical Engineering, MIT, September 1969.
28. W. H. Lee, Sampled Fourier transform hologram generated by computer, *Appl. Opt.,* **9**, No. 3, 639–643 (1970).
29. W. H. Lee, Filter design for optical data processors, *Pattern Recog.,* **2**, No. 5, 127–137 (1970).
30. C. B. Burckhardt, A simplification of Lee's method of generating holograms by computer, *Appl. Opt.,* **9**, No. 8, 1949 (1970).
31. M. A. Kronrad, N. S. Merzlyakov, and L. P. Yaroslavskii, Experiments with computer-generated holograms of transparencies, *Zh. Tekh. Fiz.,* **42**, No. 2, 414–418 (1972).
32. M. A. Kronrod, N. S. Merzlyakov, and L. P. Yaroslavskii., Experiments on digital holography, *Avtometriya,* No. 6, 30–40 (1972).
33. L. P. Yaroslavskii, M. A. Kronrod, and N. S. Merzlyakov, Computer analysis and synthesis of wavefronts, in: *Holography: Present State and Outlook for Development* [in Russian], Nauka, Leningrad (1974), pp. 54–76.
34. N. S. Merzlyakov and L. P. Yaroslavskii, Methods for recording synthesized holograms, *Abstracts of Reports, Second All-Union Conference on Holography,* Kiev, 1975 [in Russian].
35. D. Gabor, Laser speckle and its elimination, *IBM J. Res. Dev.,* **14**, 509–514 (1970).
36. W. J. Dallas, Deterministic diffusers for holography, *Appl. Opt.,* **12**, No. 6, 1179–1187 (1973).
37. J. W. Goodman, D. C. Chu, and J. R. Fienup, Recent developments in computer holograms, *Proc. SPIE,* **41**, 155–158 (1974).
38. C. N. Kurtz, H. O. Hoadley, and J. J. DePalma, Design and synthesis of random phase diffusers, *J. Opt. Soc. Am.,* **63**, No. 9, 1080–1092 (1973).
39. H. Akahori, Comparison of deterministic phase coding with random phase coding in terms of dynamic range, *Appl. Opt.,* **12**, No. 10, 2336–2343 (1973).
40. L. P. Yaroslavskii, Some approaches for visualizing information by digital holography, *Proceedings of the Conference on Computer Automatization of Research,* Novosibirsk, 1974 [in Russian].
41. L. P. Yaroslavskii and N. S. Merzlyakov, Observation of a stereoscopic effect with digital Fourier holograms, *Voprosy Radioélektroniki, Seriya Tekhnika Televideniya,* No. 3, 100–103 (1975).
42. L. P. Yaroslavskii and N. S. Merzlyakov, Observation of a stereoscopic effect with synthesized holograms, *Abstracts of Reports, Second All-Union Conference on Holography,* Kiev, 1975 [in Russian].
43. V. N. Karnaukhov, N. S. Merzlyakov, and L. P. Yaroslavskii, Three-dimensional hologram film synthesized with a digital computer, *Pis'ma Zh. Tekh. Fiz.,* **2**, No. 4, 169–172 (1976).
44. L. I. Mirkin, M. A. Rabinovich, and L. P. Yaroslavskii, Computer generation of correlated Gaussian pseudorandom numbers, *ZhVT i MF,* **5**, 1353–1357 (1972).

45. N. S. Merzlyakov and L. P. Yaroslavskii, Simulation of highlights on a diffuse surface with a programmed diffuser, *Zh. Tekh. Fiz.*, **47**, No. 6, 1263 (1977).

46. L. P. Yaroslavskii and A. M. Fayans, Study of computer processing and analysis of interferograms, in: *Iconics* [in Russian], Nauka, Moscow (1975), pp. 27–49.

47. B. P. Konstantinov, S. B. Gurevich, G. A. Gavrilov, A. A. Kolesnikov, A. B. Konstantinov, V. B. Konstantinov, A. A. Rizkin, and D. F. Chernykh, Standard facsimile channel transmission of holograms with a restricted number of half tones, *Zh. Tekh. Fiz.*, **39**, No. 2, 347–351 (1969).

48. M. A. Kronrod, N. S. Merzlyakov, and L. P. Yaroslavskii, Reconstruction of a hologram with a computer, *Zh. Tekh. Fiz.*, **42**, No. 2, 419–420 (1972).

49. L. D. Bakhrakh and A. P. Kurochkin, Use of holography in reconstruction of polar diagrams of UHF antennas from field measurements in the Fresnel zone, *Dokl. Akad. Nauk SSSR*, **171**, 1309–1312 (1966).

50. L. I. Goldfischer, Autocorrelation function and power spectral density of laser-produced speckle patterns, *J. Opt. Soc. Am.*, **55**, No. 3, 247–253 (1965).

51. L. H. Enloe, Noise-like structure in the image of diffusely reflecting objects in coherent illumination, *Bell Syst. Tech. J.*, **46**, No. 7, 1479–1489 (1967).

52. L. B. Lesem, P. M. Hirsch, Jr., and J. A. Jordan, Kinoform, *Zarubezhnaya Radio-élektronika*, No. 12, 41–50 (1969).

53. D. C. Chu, Three ways to obsolete the kinoform, *J. Opt. Soc. Am.*, **63**, 1325 (1973).

54. R. E. Haskell, Synthetic holograms and kinoforms, *Opt. Engineering*, **14**, No. 3, 195–199 (1975).

55. L. P. Yaroslavskii, *Image Input and Output Apparatus for Digital Computers* [in Russian], Énergiya, Moscow (1968).

56. M. P. Grishin, Sh. M. Kurbanov, and V. P. Markelov, *Automatic Computer Entry and Processing of Photographic Images* [in Russian], Énergiya, Moscow (1976).

57. T. Sandor and G. Cagliusio, Rotating drum scanner-display system for digital image processing, *Rev. Sci. Instrum.*, **45**, No. 4, 506–509 (1974).

58. L. B. Lesem, P. M. Hirsch, and J. A. Jordan, Computer synthesis of holograms for 3-D display, *Commun. of Associat. for Computing Machinery*, **11**, 661–674 (1968).

59. G. W. Stroke, M. Halioua, and V. Strinivasan, Holographic image restoration using Fourier spectrum analysis of blurred photographs and computer aided synthesis of Wiener filters, *Phys. Lett.*, **51A**, No. 7, 383–385 (1975).

60. T. Yatagai, Stereoscopic approach to 3-D display using computer-generated holograms, *Appl. Opt.*, **15**, No. 11, 2722–2729 (1976).

61. N. S. Merzlyakov and L. P. Yaroslavskii, Image visualization by means of computer-synthesized holograms, *Dokl. Akad. Nauk SSSR*, **237**, No. 2, 318–321 (1977).

62. L. P. Jaroslavski and N. S. Merzlyakov, Stereoscopic approach to 3-D display using computer-generated holograms: comment, *Appl. Opt.*, **16**, No. 8, 2043 (1977).

63. L. P. Jaroslavski and N. S. Merzlyakov, Information display using the methods of digital holography, *Computer Graphics and Image Processing*, **10**, 1–29 (1979).

64. V. N. Karnaukhov and N. S. Merzlyakov, Holographic three-dimensional kinoform film synthesized by computer, *Third All-Union Conference on Holography*, Ul'yanovsk, 1978 [in Russian].

65. N. R. Popova and L. P. Yaroslavskii, Digital model for studying speckle noise in holography, *All-Union Scientific and Technological Conference on Automation of Experimental Research*, Kuibyshev, 1978 [in Russian], p. 158.

66. R. J. Collier, C. B. Burckhardt, and L. H. Lin, *Optical Holography*, Academic Press, New York (1972).